PATTERSON FAMILY HISTORY

Compiled by

Betty Jewell Durbin Carson
DAR Member #832584

for

Dorothy A. Patterson

HERITAGE BOOKS
2015

HERITAGE BOOKS
AN IMPRINT OF HERITAGE BOOKS, INC.

Books, CDs, and more—Worldwide

For our listing of thousands of titles see our website
at
www.HeritageBooks.com

Published 2015 by
HERITAGE BOOKS, INC.
Publishing Division
5810 Ruatan Street
Berwyn Heights, Md. 20740

Heritage Books by the author:

The Brice Family Who Settled in Fairfield County, South Carolina, about 1785 and Related Families

Durbin and Logsdon Genealogy with Related Families, 1626–1991

The Durbin and Logsdon Genealogy with Related Families, 1626–1991, Volume 2

Durbin and Logsdon Genealogy with Related Families, 1626–1994

Durbin and Logsdon Genealogy with Related Families, 1626–1998

Durbin-Logsdon Genealogy and Related Families from Maryland to Kentucky, Volumes 1–2

CD: The Durbin and Logsdon Genealogy with Related Families, 1626–2000, 3rd Revised Edition

History of Curtis Land, 1635–1683; with Excerpt on Francis Land

Our Ewing Heritage, with Related Families, Part One and Two, Revised Edition
Betty Jewell Durbin Carson and Doris M. Durbin Wooley

CD: Our Ewing Heritage, with Related Families, Revised Edition
Betty Jewell Durbin Carson and Doris M. Durbin Wooley

Patterson Family History

International Standard Book Numbers
Paperbound: 978-0-7884-5600-8
Clothbound: 978-0-7884-6073-9

PATTERSON

FAMILY

HISTORY

Compiled by

Betty J. Carson, DAR Member
832584

For

Dorothy A. Patterson

October, 2014

DEDICATION

As of today, only the descendants remain of the eighteen **Patterson** children. Information is needed and must be share among ourselves. Please forward any **Patterson** pictures, names, stories, or corrections to: Cynthia Chappelle (803) 260-0093 or Dorothy Chappelle (864) 374-5322.

This book is only the beginning, so please help with putting together the next **"Patterson Family"** book by providing information and pictures. Let's keep the family history going.

Be Blessed and Thank You!

Special thanks to Mrs. Betty J. Carson who tirelessly assisted in the Patterson Family research. Also, special thanks to Dorothy Chappelle for always having family on her mind and in her heart.

Another big special thanks goes to Wilma Kirkland who got Dorothy interested in the family genealogy. Without her encouragement and help, this book would not have been possible.

TABLE OF CONTENTS

PATTERSON HISTORY

"4426 Cokesbury Road, Hodges, SC 29653
June 18, 2009

My grandmother was born in the 1800's. She was married to Rube or Ruben Patterson in the year 1900 (census shows 1899). Her name was Willie Ann Holloway and her father's name was Sam Holloway. Ruben's parents: Willis and Nancy McKensey. Siblings: Metz, Otis, Tom, Johnny, Willie, Elbert, James, Sam, Ruben, Wince, Chaney, Nelson, Jane, Caroline, George, Ben; it is believed to have been many more (approx. 18).

Her mother was Martha, from Queensland, Australia. She had sisters Emma, Mary, Martha, Caroline, Ada, Susa, and Adeline; she had more but I just don't have all of their names. Her brothers: John, Henry, Matthew, Morris, West, Sam, George, Edward, and Solan. It was a big family of them.

Their Daddy was a slave and named Sam after his father. His mother was a slave, also.

Letter from Dorothy A. Chappelle"

Willis Patterson & Nancy McKenzie Patterson

Willis Patterson's name was changed from Joe Crews to Pink Patterson; however, once he became a free man, he changed his name to Willis Patterson. Willis was a full-blooded Indian who was sold to his master by Joe Crews, the biggest slave trader in the country. He was stolen when a young buck somewhere in Mississippi, along with other Indians, and sold into slavery with the "niggers." After he was purchased by Mr. Joe Patterson, his name became "Pink Patterson."

Nancy McKenzie (McKensey) was a white woman who came from Ireland and was working on the Patterson farm. She was not a slave, but was married to Willis Patterson, his father, by his "Marster." Nancy was born in 1841, died June 20, 1939, age 98, buried Geeenville, South Carolina. Her death certificate gives her parents as Millie and Perry Cunningham, Route 2, Greenville, SC; buried Lonnie Hill Cemetery, undertaker Biggs Stewart, Inc., Greenville, South Carolina, June 22 1939. Another researcher states her father was white and came from Ireland or Scotland. Perhaps McKenzie stems from McCants/McKantz. See History of Williamsburg, SC later in book.

Children of Willis Patterson and Nancy McKenzie:

1. Metz--see 1940 Census McCormick, Mccormick, SC below
2. Otis—see 1940 Census, McCormick, McCormick, South Carolina:
3. Tom --see 1930 Bordeaux, McCormick, SC census
4. Johnny--see 1930 Atlanta, Fulton, GA census below
5. Willie-- see 1940 Census, McCormick, McCormick, SC

6. Elbert-see 1940 McCormick, McCormick, SC census below
7. James-see Jim Patterson on 1870 Lexington, SC census below
8. Sam-see 1940 McCormick, McCormick, SC census below
9. Ruben-see Ruben on 1940 Greenwood, SC census below
10. Wince-see 1920 census, Fellowship, Greenwood, SC below
11. Chaney
12. Jane-living with brother, Wince & Ida, 1900 Census, Fellowship, Greenwood, SC
13. Caroline-- see 1870 Census, Ninety-Six, Abbeville, SC below
14. George-see 1880 Census, Calhoun Mills, Abbeville, SC below
15. Nelson-see 1870 Census, Ninety-Six, Abbeville & Lexington, SC below
16. Richard
17. Washington
18. Jesse--see 1870 Census, Ninety-Six, SC below

The following family members have been compiled from birth and census records. It must be remembered that Edgefield, Newberry, and Spartanburg Counties were once part of the Abbeville District. Before 1860, Negroes were listed (if listed) as Indians. It was 1870 and later before they were classified as Blacks or Mulattoes.

1870 Census, Ninety-Six, Abbeville, South Carolina:

Jesse Patterson	Male	B	35	Farm Laborer	Born South Carolina
Caroline Patterson	Female	B	25	Farm Laborer	Born South Carolina
Nelson Patterson	Male	B	32	Farm Laborer	Born South Carolina

1870 Census, Lexington, Lexington Courthouse, South Carolina:

Jim Patterson	Male	B	34	Farm Laborer	Born South Carolina
Sylvia Patterson, wife	Female	B	30	Farm Laborer	Born South Carolina
Joshua Patterson	Male	B	14	Farm Laborer	Born South Carolina
Nelson Patterson	Male	B	12	Farm Laborer	Born South Carolina
Furman Patterson	Male	B	6		Born South Carolina
Dinah Patterson	Female	B	4		Born South Carolina

1880 Census, Calhoun Mills, Abbeville, South Carolina:

George Patterson	Male	B	40	
Ester Patterson	Female	B	35	Wife
Gilbert Patterson	Male	B	12	
Charity Patterson	Female	B	10	
Wade Patterson	Male	B	7	
Robert Patterson	Male	B	6	
Jane Patterson	Female	B	2	

1880 Census, Hibler, Edgefield, South Carolina:

George Patterson	Male	B	42
Rachael Patterson	Female	B	40
Seymour Patterson	Male	B	12

On same page:

Polk Hollaway	Male	B	34	
Susan Hollaway, wife	Female	B	27	
Harriet Hollaway	Female	B	7	
Henry Hollaway	Male	B	2	
Infant Daughter	Female	B	3/12	

Delayed birth certificate for Will Patterson (Will Moore?), born Feb. 25, 1888, parents given as Reuben Patterson and Willie Ann Hollaway.

1900 Census, Society Hill, Darlington, South Carolina:

Cornieulius Patterson	Male	B	43	b. Sep. 1856
Ida Patterson	Female	B	37	Wife
Loucracia Patterson	Female	B	18	
Winston Patterson	Male	B	12	
Benjamin Patterson	Male	B	10	
William Patterson	Male	B	8	
Carrie Patterson	Female	B	7	
Mabel Patterson	Female	B	5	
Cornelious Patterson	Male	B	2	
Liza A. Patterson	Female	B	8/12	

1900 Census, Fellowship, Greenwood, South Carolina:

Wince Patterson	Male	B	25	
Ida Patterson, wife	Female	B	21, married 1899	
Janie Patterson	Female	B	18	Sister

1900 Census, Sycamore, Barnwell, South Carolina:

Willis Patterson	Male	B	30	Born May 1892, SC
Phoeby, wife	Female	B	23	
Arthur, son	Male	B	8	
Richard Patterson	Male	B	25	Brother
Washington Patterson	Male	B	20	Brother

1900 Census, Bordeaux, Abbeville, South Carolina:

Wm Patterson	Male	B	30	Farm Laborer, Born South Carolina
Manie, wife	Female	B	28	Keeping House
Allen, son	Male	B	8	
George, son	Male	B	6	
Permilla, dau	Female	B	2	
Francis, son	Male	B	2/12	
Penter Dalphus	Female	B	22	Servant

1900 Census, Fellowship, Greenwood, South Carolina:

Rube Patterson	Male	B	50

Willy Ann Patterson	Female	B	25	
George Patterson	Male	B	6 1/12	
Will Moore	Male	B	1 1/12	

1910 Census, Society Hill, Darlington, South Carolina:

C. R. Patterson	Male	B	51	b. Sep 1856 (Cornelius)
Ida E. Patterson	Female	B	44	Wife
Winston Patterson	Male	B	21	
Alvarine T. Patterson	Female	B	20	
William F. Patterson	Male	B	17	
Carrie Patterson	Female	B	15	
Mabel Patterson	Female	B	12	
C. R. Patterson, Jr.	Male	B	11	
Eliza A. Patterson	Female	B	10	
Outlaw Patterson	Male	B	7	

1910 Census, Greenwood, Greenwood, South Carolina:

Rube Patterson	Male	B	32	
Willie Ann Patterson	Female	B	30	
Carrie Patterson	Female	B	9	
Mamie Patterson	Female	B	7	
Willie Patterson	Male	B	5	
Mose Patterson	Male	B	4	
Ruben Patterson	Male	B	2	
Ebbie Patterson		B	3/12	

On same page:

Reuben Holloway	Male	B	24	
Emma Holloway	Female	B	23	Wife
Evander Patterson	Male	B	4	
Annie Joe Patterson	Female	B	2	

1920 Census, Cokesbury, Greenwood, South Carolina:

Reuben Patterson	Male	B	44	
W. A. (Willie Ann)	Female	B	40	Wife
Louise Patterson	Female	B	18	Daughter
M. F. (Mamie L.)	Female	B	14	Daughter
W. R. (Willie Robert)	Male	B	14	Son
Rock (David)	Male	B	13	Son
Reuben Patterson	Male	B	10	Son
T. L. (Robert)	Male	B	9	Son
L. S. (Lewis)	Male	B	7	Son
W. M. (Willie Mae)	Female	B	5	Daughter
Edward Patterson	Male	B	2	Son
Infant		B	4/12	
Elva Patterson	Female	B	6/12	Daughter Glenen
Malik Jones	Female	B	13	Niece Mae Bell

1920 Census, Township 2, Fairfield, South Carolina:

Samuel Johnson	Male	B	25	b. South Carolina
Janie Johnson	Female	B	24	b. South Carolina
Jessie Johnson	Female	B	2 8/12	
Alberta Johnson	Female	B	0 6/12	

1920 Census, Walnut Grove, Greenwood, South Carolina:

Samie Johnson	Male	B	52
Jane Johnson	Female	B	48
Susie Johnson	Female	B	24
Mary Johnson	Female	B	22
John Johnson	Male	B	20
Hiram Johnson	Male	B	18
Willie Johnson	Male	B	16
Mary Johnson	Female	B	14
Lillie Johnson	Female	B	12
Clara Johnson	Female	B	10

1920 Census, Mont Clare, Darlington, South Carolina:

Winston Patterson	Male	B	35	
Meda Patterson	Female	B	30	
James Patterson	Male	B	4	
Willis Patterson	Male	B	60	Father-in-law

1920 Census, Fellowship, Greenwood, South Carolina:

Wence Patterson	Male	B	38, General Farm, rented
Ida, wife	Female	B	35
Fredie, son	Male	B	17
Annie, dau	Female	B	15
Rosa, dau	Female	B	11
Pearl, dau	Female	B	9
Hemy, son	Male	B	5
Frances, dau	Female	B	4
Carry, dau	Female	B	

1930 Census, Manhattan, New York, New York:

Samuel Johnson	Male	B	36	Porter in private home
Jennie (Janie)	Female	B	30	Wife
Helen Johnson	Female	B	1 10/12	
John D. Bradford	Male	B	32	
Alberta Bradford	Female	B	25	Daughter of Samuel & Janie
Harry Brown	Male	B	21	Lodger

PATTERSON FAMILY

1930 Census, Greenwood, Greenwood, South Carolina:

Reuben Patterson	Male	B	65	
Willie A. Patterson	Female	B	50	Wife
David Patterson	Male	B	24	Son
Reuben Patterson	Male	B	21	Son
T. L. Patterson	Male	B	20	Son
L. F. Patterson	Male	B	18	Son
William F. Patterson	Male	B	15	Son
Eddie Patterson	Male	B	12	Son
Evelyn Patterson	Female	B	10	Daughter
Cora I. Patterson	Female	B	7	Daughter
Kathleen Patterson	Female	B	10	Daughter (twin to Evelyn?)
May B. Jones	Female	B	24	Niece

Willie Ann Patterson is buried at Beulah Baptist Cemetery, Greenwood, South Carolina. Her sister, Mary Hollway Williams is buried at Mt. Tabor Church Cemetery, Greenwood, South Carolina.

1930 Census, Atlanta, Fulton, Georgia:

Rich Patterson	Male	B	59	
Lula, wife	Female	B	55	
Johnny Patterson	Male	B	39	Brother
Emma Patterson	Female	B	30	Johnny's wife
AnnaL.	Female	B	13	Dau of Johnny & Emma
Billie Dickson	Male	B	28	
Rosa Dickson	Female	B	26	

1930 Census, Bordeaux, McCormick, South Carolina:

Tom Patterson	Male	B	35	
Adeline Patterson	Female	B	74	Mother
John Patterson	Male	B	38	Brother
Mae Patterson	Female	B	38	Sister-in-law (John's wife)
Metz Patterson	Male	B	23	Brother
Ethel Patterson	Female	B	23	Sister-in-law (Metz' wife)

1930 Census, Troy, Greenwood, South Carolina:

Willie Patterson	Male	B	23	b. South Carolina
Eveline Patterson	Female	B	27	b. South Carolina
Lily Maude Patterson	Female	B	11	
Louise Patterson	Female	B	10	
Ida May Patterson	Female	B	6	
Daisy Belle Patterson	Female	B	46112	
Willie Patterson	Male	B	2 7112	
Janie Lee Patterson	Female	B	0 4/12	
Mamie Lou Kennedy	Female	B	10	Grandaughter
Beatrice Kennedy	Female	B	9	Grandaughter

1930 Census, Fort Lauderdale, Broward County, Florida:

Wince Patterson	Male	B	48	b. South Carolina
Ida Patterson, wife	Female	B	46	b. South Carolina
Rosa Patterson	Female	B	18	
Pearl Patterson	Female	B	16	
Heneryetta Patterson	Female	B	14	
Frances Patterson	Female	B	13	
Carrie Patterson	Female	B	10	
Ruben Patterson	Male	B	9	

1940 Census, McCormick, McCormick, Edgefield, South Carolina:

Elbert Patterson	Male	B	45	Laborer, elementary 3rd grade
Gladys	Female	B	24	
Marian	Female	B	19	
Alberta	Female	B	19	
Tona	Female	B	13	
Charlie Mae	Female	B	8	
Virginia	Female	B	2	
Lizzie Kimbrell	Female	B	20	Sister-in-law, servant

1940 Census, McCormick, McCormick, South Carolina:

Otis Patterson	Male	B	38, Laborer, rented, Elementary, 4th grade	
Anna, wife	Female	B	32	
Louise, dau	Female	B	16	
Lillie Mae, dau	Female	B	13	1930 census lists as Ella M. Patterson
Kate, dau	Female	B	11	
Otis, son	Male	B	9	
Rosa Lee, dau	Female	B	6	
Eddie Ruth, dau	Female	B	3	

1940 Census, Greenwood, South Carolina:

Ruben Patterson	Male	B	77	Elementary School, 1st grade b. SC Farm Laborer
Willie Anne, wife	Female	B	62	Keeping House
Ellen S., Dau	Female	B	20	Evelyn (Evie or Evelynie)
Kathleen, Dau	Female	B	19	
Cora Emma, Dau	Female	B	17	
Willie May, Dau	Female	B	25	
Corrine, Gdau	Female	B	4	
Dorothy Anne, Gdau	Female	B	2	
Eddie L., Gson	Male	B	0	
Mary Belle, Dau	Female	B	34	

1940 Census, McCormick, McCormick, South Carolina:

Sam Patterson	Male	B	49,	Delivery man, elementary, 3rd grade

Laura, wife	Female	B	50	
Geneva Cantelou	Female	B	23	Servant

1940 Census, McCormick, McCormick, South Carolina:

Willie Patterson	Male	B	59	Laborer Coal Chute, Born SC
Sarah, wife	Female	B	60	
Clifton, son	Male	B	27	
Cornelia, dau	Female	B	16	

On the 1880 District 85, Jefferson, Georgia is listed:

Robert Patterson		Male	B	50	Laborer	Born GA, Mother/Father b. GA
Charlott	Wife	Female	B	50	Laborer	Born GA, Mother/Father b. GA
Hester	Dau	Female	B	20	Laborer	Born GA, Mother/Father b. GA
Elvira	Dau	Female	B	18	Laborer	Born GA, Mother/Father b. GA
Sam	Son	Male	B	10	Laborer	Born GA, Mother/Father b. GA
Ellen	Dau	Female	B	8		Born GA, Mother/Father b. GA
Mary	Dau	Female	B	6		Born GA, Mother/Father b. GA
Jane	Dau	Female	B	2		Born GA, Mother/Father born GA

On the 1880 District 1068, Spalding, Georgia is listed:

Rebecca Patterson	Widow	Female	B	40	Keeping house	Born GA, Mother/Father born GA
Willis Patterson	Son	Male	B	21	Farm Worker	Born GA, Mother/Father born GA
Thomas Patterson	Son	Male	B	18	Farm Worker	Born GA, Mother/Father born GA
Joseph Patterson	Son	Male	B	15	Farm Worker	Born GA, Mother/Father born GA
Maddison Patterson	Son	Male	B	13	Farm Worker	Born GA, Mother/Father born GA
Mariah Patterson	Dau	Female	B	12		Born GA, Mother/Father born GA
Louisa Patterson	Gdau	Female	B	10		Born GA, Mother/Father born GA

On the 1880 District 46, District 1173, Mitchell, Georgia is listed:

Willis Patterson		Male	B	38	Farm Laborer	Born GA, Mother/Father born GA
Jane Patterson	Wife	Female	B	35	Keeping House	Born GA, Mother/Father born GA
Rubbin Rhodes	SSon	Male	B	15	Farm Laborer	Born GA, Mother/Father born GA
Tilman Patterson	Son	Male	B	10	Farm Laborer	Born GA, Mother/Father born GA
Annie Patterson	Dau	Female	B	8		Born GA, Mother/Father born GA
Carrie Patterson	Dau	Female	B	6		Born GA, Mother/Father born GA
Allen Patterson	Son	Male	B	4		Born GA, Mother/Father born GA
Emma Patterson	Dau	Female	B	3		Born GA, Mother/Father born GA
Catherine Patterson	Dau	Female	B	2		Born GA, Mother/Father born GA

Grave Search Results

Beaulah Baptist Church, Greenwood, Greenwood, South Carolina:

McBride, Leona Wells	b. Apr. 18, 1902, d. Nov. 12, 1986	103632522
McBride, Luella	b. 1901, d. 1959	103632614
McBride, Rosie H.	b. Mar. 9, 1891, d. Jul. 30, 1981	103632510
McBride, Sadie	b. 1922, d. 2011	103632288
Nedward, Willis W. (Kathleen Brook)	b. Mar. 15, 1871, d. Apr. 29, 1955	103632126

Patterson, David	b. 1910,d. 1950	103631781
Patterson, Mae B.	b. 1914,d. 1954	103632550
Patterson, Ruben	b. unknown, d. Oct. 15, 1941	103631768
Patterson, Willie Ann	b.Nov. 25, 1873, d. Oct. 12, 1947	103632637

Newmansville African American Cemetery, Alachua, Alachua Co., **Florida:**

Johnson, Janie	b. Apr. 6, 1872, d. Jul. 4, 1970	124896162

Added by
Thomas George

Johnson, Samuel	b.Feb.5, 1896,d.Aug. 16,1975	100449239

(buried Saint Pauls Memorial Cemetery, Alachua, Alachua, Florida)

Added by
Emery, 1959

Evergreen Cemetery, Fort Lauderdale, Broward County, Florida:

Patterson, Earlie	b. 1890, d. 1946	123979995
Patterson, Emma	b. unknown, d. unknown	123979996
Patterson, George Denison	b. Sep. 28, 1906, d. Apr. 21, 1986	18824762

Forest Lawn Memorial Gardens, Fort Lauderdale, Broward County, Florida:

Patterson, Robert Lee	b. Mar. 4, 1952, d. Apr. 5, 2009	35885905
Patterson, Algreen Seth	b. Oct. 20, 1972, d. Jan. 6, 1995	99406516
Patterson, Hortense	b. 1940, d. 2013	120856594
Patterson, Katherine D.	b. 1913, d. 1980	63646898

PATTERSON FAMILY

Our Lady Queen of Heaven, Fort Lauderdale, Broward County, Florida:

Patterson, Allen W.	b. 1928, d. 2005	14646136
Patterson, Ruth	b. 1928, d. 2005	14646142

Sunset Memorial Gardens, Fort Lauderdale, Broward County, Florida:

Patterson,Elsie	b.Jul.20, 1915,d.Feb.17,2012	105129217
Patterson, Floyd	b. Oct. 31, 1961, d. Apr. 13, 1991	96772602
Patterson, Toni	b. Apr. 4, 1977, d. Dec. 8, 2011	14646142

Lauderdale Memorial Park, Fort Lauderdale, Broward County, Florida:

Patterson, Catherine L.	b. Aug. 19, 1918, d. Mar. 24,2010	75862596
Patterson, Earlie	b. 1890, d. 1946	123979995
Patterson,Edgar L.	b. Sep.23, 1909,d.Jan.17, 1988	75862080
Patterson, Elinore R.	b. 1913, d. 1992	75856698
Patterson, Geneva	b. 1909, d. 1972	72426651
Patterson, William C.	b. 1907, d. 1989	7556653
Patterson, Patricia M.	b. May 18, 1935, d. Feb. 10, 1973	97858654
Patterson, Raymond G.	b. 1907, d. 1973	72426553

Woodlawn Cemetery, Fort Lauderdale, Broward County, Florida:

Patterson, Wince	b. 1875, d. 1939	18249210
Patterson, Ida May	b. 1871, d. 1951	31117934

(also, given as burial in Dania Memorial Park, Fort Lauderdale, Broward County, Florida, By GerbLady, Plot: Section E, Lot 8 67840015)

Patterson, William J.	b. Jul. 23, 1872, d. Apr. 20, 1960	67840016

(given as burial in Dania Memorial Park, Fort Lauderdale, Broward County, Florida, 678400 15)

Patterson, Pvt. Eddie Lloyd	b. May 3, 1931, d. Oct. 3, 1977	100241158 (US Army, Korea)
Patterson, Unknown	b. unknown, d. 1995	106725950

(cemetery photos added by Mary Harrell-Sesniak, daughter of Wince & Ida May Patterson; Pvt. Eddie Lloyd Patterson, son or grandson.)

1870 Scuffletown, Laurens County, South Carolina Census:

Patterson, Pinkney	Male	B	46
Patterson, Mary	Female	B	38
Patterson, George	Male	B	17
Patterson, Pinkney	Male	B	15
Patterson, Benjamin	Male	B	13

Patterson, William	Male	B	21
Patterson, Jane	Female	B	30
Patterson, Bluford	Male	B	8
Patterson, Benjamin	Male	B	6
Patterson, Martha	Female	B	3
Patterson, Lenoir	Female	B	1
Patterson, Jessey	Male	B	2/12

1870 Great Cypress, Barnwell County, South Carolina Census:

Patterson, Pink	Male	B	29
Patterson, Sarah	Female	B	29
Patterson, Sarah	Female	B	8
Patterson, Rosa	Female	B	7
Patterson, John	Male	B	6
Patterson, Billie	Male	B	5
Patterson, Prinius	Male	B	3
Patterson, Pink	Male	B	8/12

DEATH CERTICATES

N. B.—WRITE PLAINLY, WITH UNFADING INK—THIS IS A PERMANENT RECORD. Every item of information should be carefully supplied. AGE should be stated EXACTLY. PHYSICIANS should state CAUSE OF DEATH in plain terms, so that it may be properly classified. Exact statement of OCCUPATION is very important. See instructions on back of certificate.

MARGIN RESERVED FOR BINDING

Standard Certificate of Death
STATE OF SOUTH CAROLINA
Bureau of Vital Statistics
State Board of Health

File No.—For State Registrar Only: 17741

1. PLACE OF DEATH
County of Greenwood
Township of Greenwood
or
City of Greenwood (No. R.F.D. 4 St.; Box 77 Ward) (If death occurred in a Hospital or institution give its NAME instead of street and number.)
Registration District No. 2306
Registered No. 46 (For use of Local Registrar)
Home Address Route 4 Box 77.
Residence— In City.......Yrs.......Mos.......Days

2. FULL NAME Ruben Patterson

PERSONAL AND STATISTICAL PARTICULARS

3. SEX: male
4. COLOR OR RACE: Col'd
5. Single, Married, Widowed, or Divorced (write the word): married
5a. If married, widowed, or divorced HUSBAND of (or) WIFE of: Willie Ann Patterson
6. DATE OF BIRTH (Month, day, and year): 1881
7. AGE: Years 60 | Months | Days | If less than 1 day, ...hrs. or ...min.

OCCUPATION
8. Trade, profession, or particular kind of work done, as spinner, sawyer, bookkeeper, etc.: Common Labor
9. Industry or business in which work was done, as silk mill, saw mill, bank, etc.: Farmer
10. Date deceased last worked at this occupation (month and year) ....
11. Total time (years) spent in this occupation ....

12. BIRTHPLACE (city or town) (State or Country): Greenwood, " County

FATHER
13. NAME: Willie Patterson
14. BIRTHPLACE (city or town) (State or Country): Greenwood, " County

MOTHER
15. MAIDEN NAME: Nancy McKenzie
16. BIRTHPLACE (city or town) (State or Country): Greenwood, " County

17. INFORMANT: Nelson Patterson
(Address) Greenwood, R4 B.
18. BURIAL, CREMATION, OR REMOVAL
Place: [illegible] Cemetery 10/17/41
19. UNDERTAKER: Robinson & Son
(Address) [illegible]
20. FILED Dec 18, 1941 Julia Lee, Registrar.

MEDICAL CERTIFICATE OF DEATH

21. DATE OF DEATH (month, day, and year) 10/15/1941
22. I HEREBY CERTIFY, That I attended deceased from 9/8/41, 19..., to 10/15/41, 19...
I last saw him alive on 10/12/41, 19..., death is said to have occurred on the date stated above, at 3 A. m.
The principal cause of death and related causes of importance in order of onset were as follows: Chronic Myocarditis, Nephritis | Date of onset

Was this death due to pregnancy or to childbirth? If so, state which 131-2

Contributory causes of importance not related to principal cause: ....

Name of operation........ Date of........
What test confirmed diagnosis?........ Was there an autopsy?........
23. If death was due to external causes (violence) fill in also the following:
Accident, suicide, or homicide?........ Date of injury........, 19...
Where did injury occur?........ (Specify city or town, and state)
Specify whether injury occurred in industry, in home, or in public place.
Manner of injury........
Nature of injury........
Was disease or injury in any way related to occupation of deceased?........
If so, specify........
(Signed) R. B. [illegible], M. D.
(Address) Greenwood, S.C.

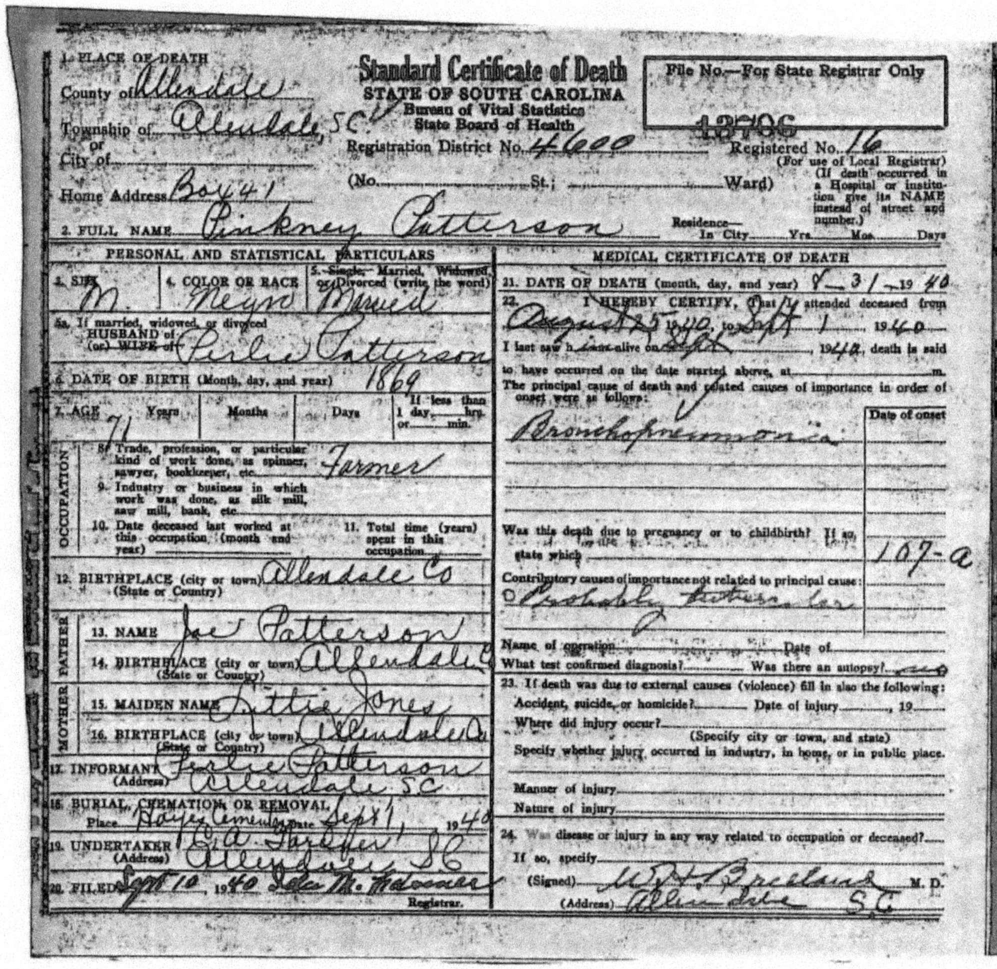

Standard Certificate of Death
STATE OF SOUTH CAROLINA
Bureau of Vital Statistics
State Board of Health

File No.—For State Registrar Only
13706

1. PLACE OF DEATH
County of Allendale
Township of Allendale S.C.
or City of
Registration District No. 4600
Registered No. 16
(For use of Local Registrar)
Home Address Box 41
(No. St.; Ward)
(If death occurred in a Hospital or institution give its NAME instead of street and number.)

2. FULL NAME Pinkney Patterson
Residence—In City Yrs. Mos. Days

PERSONAL AND STATISTICAL PARTICULARS

3. SEX M.
4. COLOR OR RACE Negro
5. Single, Married, Widowed, or Divorced (write the word) Married
5a. If married, widowed, or divorced HUSBAND of (or) WIFE of Perlie Patterson
6. DATE OF BIRTH (Month, day, and year) 1869
7. AGE 71 Years Months Days If less than 1 day, hrs. or min.

OCCUPATION
8. Trade, profession, or particular kind of work done, as spinner, sawyer, bookkeeper, etc. Farmer
9. Industry or business in which work was done, as silk mill, saw mill, bank, etc.
10. Date deceased last worked at this occupation (month and year)
11. Total time (years) spent in this occupation

12. BIRTHPLACE (city or town) (State or Country) Allendale Co

FATHER
13. NAME Joe Patterson
14. BIRTHPLACE (city or town) (State or Country) Allendale C

MOTHER
15. MAIDEN NAME Littie Jones
16. BIRTHPLACE (city or town) (State or Country) Allendale Co

17. INFORMANT Perlie Patterson
(Address) Allendale S.C.

18. BURIAL, CREMATION, OR REMOVAL
Place Hayes Cemetery Date Sept 1, 1940

19. UNDERTAKER R. A. Lawton
(Address) Allendale, S.C.

20. FILED Sept 10, 1940 Ida M. Widener
Registrar.

MEDICAL CERTIFICATE OF DEATH

21. DATE OF DEATH (month, day, and year) 8-31-1940

22. I HEREBY CERTIFY, That I attended deceased from August 25 1940, to Sept 1 1940
I last saw h im alive on Sept , 1940, death is said to have occurred on the date stated above, at m.
The principal cause of death and related causes of importance in order of onset were as follows:
Date of onset
Bronchopneumonia

Was this death due to pregnancy or to childbirth? If so, state which
107-a

Contributory causes of importance not related to principal cause:
Probably tuberculosis

Name of operation Date of
What test confirmed diagnosis? Was there an autopsy? no

23. If death was due to external causes (violence) fill in also the following:
Accident, suicide, or homicide? Date of injury , 19
Where did injury occur?
(Specify city or town, and state)
Specify whether injury occurred in industry, in home, or in public place.
Manner of injury
Nature of injury

24. Was disease or injury in any way related to occupation or deceased?
If so, specify

(Signed) W. H. Breeland M. D.
(Address) Allendale S.C.

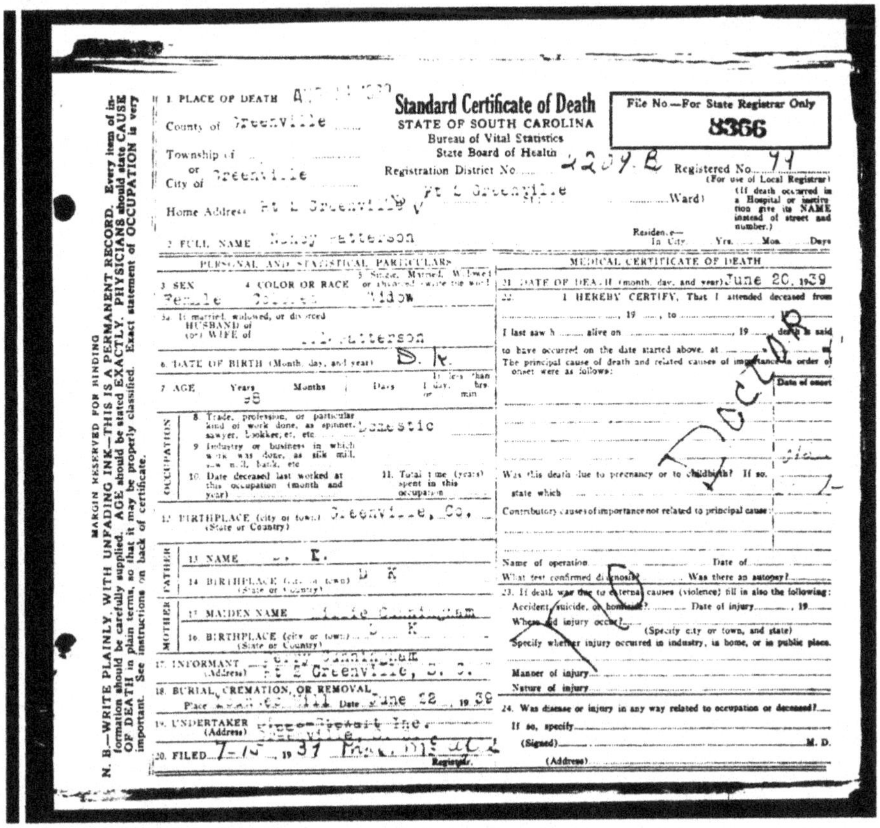

Standard Certificate of Death
STATE OF SOUTH CAROLINA
Bureau of Vital Statistics
State Board of Health

File No.—For State Registrar Only 8366

1. PLACE OF DEATH
County of Greenville
Township of ... or City of Greenville
Registration District No. 2209 B
Registered No. 71
2. FULL NAME ... Patterson
PERSONAL AND STATISTICAL PARTICULARS
3. SEX Female
4. COLOR OR RACE Colored
5. Single, Married, Widowed or Divorced Widow
6. DATE OF BIRTH D. K.
7. AGE 98
8. Trade, profession, or particular kind of work done Domestic
12. BIRTHPLACE (city or town) Greenville, Co.
13. NAME (of father) D. K.
14. BIRTHPLACE D K
16. BIRTHPLACE D K
17. INFORMANT
(Address) Greenville, S. C.
18. BURIAL, CREMATION, OR REMOVAL
Date June 22, 1939
19. UNDERTAKER
20. FILED 7-15 1939
MEDICAL CERTIFICATE OF DEATH
21. DATE OF DEATH (month, day, and year) June 20, 1939
I HEREBY CERTIFY, That I attended deceased from
The principal cause of death and related causes of importance in order of onset were as follows:
Was this death due to pregnancy or to childbirth? If so, state which
Contributory causes of importance not related to principal cause:
Name of operation ... Date of ...
What test confirmed diagnosis ... Was there an autopsy?
23. If death was due to external causes (violence) fill in also the following:
Accident, suicide, or homicide? ... Date of injury ... 19...
Where did injury occur? (Specify city or town, and state)
Specify whether injury occurred in industry, in home, or in public place.
Manner of injury
Nature of injury
24. Was disease or injury in any way related to occupation or deceased?
If so, specify
(Signed) ... M. D.
(Address)

N. B.—WRITE PLAINLY, WITH UNFADING INK—THIS IS A PERMANENT RECORD. Every item of information should be carefully supplied. AGE should be stated EXACTLY. PHYSICIANS should state CAUSE OF DEATH in plain terms, so that it may be properly classified. Exact statement of OCCUPATION is very important. See instructions on back of certificate.

1870 Census, Township 13, Fairfield, South Carolina:

Cornelius Patterson	Male	B	50	
Frances Patterson	Female	B	45	
Mary Jackson	Female	B	24	Daughter?
Nathan Jackson	Male	B	2	
John Jackson	Male	B	6/12	
Emma Jackson	Female	B	8	
Robert Mans	Male	B	28	Boarder

1880Census, Township 13, Fairfield, South Carolina:

Cornelius Patterson	Male	Mulatto	60	
Silvia Patterson	Female	B	38	Wife
Gennie Patterson	Female	B	13	
Christina Patterson	Female	B	11	
Robert Patterson	Male	B	8	
Warren Patterson	Male	B	6	

1930 Census, Darlington, Darlington, South Carolina:

Willie Patterson	Male	B	36	born South Carolina
Nancy Patterson	Female	B	26	
Ida E. Patterson	Female	B	7	
Joe Patterson	Male	B	5	
Outlaw Patterson	Male	B	26	Brother

Ance try South Carolina, Death Recorda, 1821-1960 for Outlaw Pattel'8on

1950-1961 > 1954 > Florence

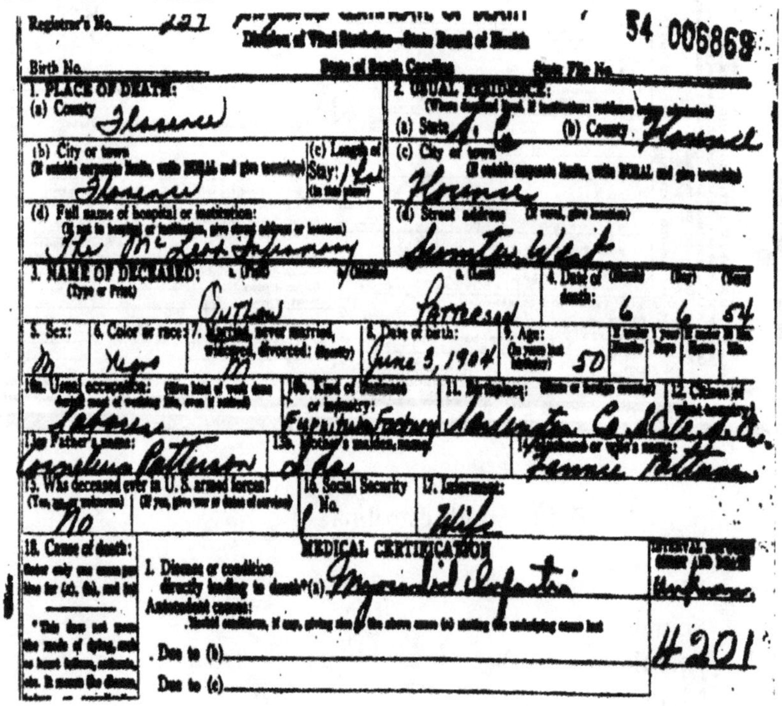
Registrar's No. 54 006868
Division of Vital Statistics—State Board of Health
State of South Carolina
1. PLACE OF DEATH: (a) County Florence
(b) City or town Florence
(d) Full name of hospital or institution: The McLeod Infirmary
2. USUAL RESIDENCE: (a) State S. C. (b) County Florence
(c) City or town Florence
(d) Street address Sumter West
3. NAME OF DECEASED: Outlaw Patterson
4. Date of death: 6 6 54
5. Sex: M 6. Color or race: Negro 7. Married, never married, widowed, divorced: M
8. Date of birth: June 5, 1904 9. Age: 50
10a. Usual occupation: Laborer 10b. Kind of business or industry: Furniture Factory
11. Birthplace: Darlington Co. S.C.
13. Father's name: Cornelius Patterson 13b. Mother's maiden name: Ida
14. Husband or wife's name: Fannie Patterson
15. Was deceased ever in U. S. armed forces? No
16. Social Security No.
17. Informant: Wife
18. Cause of death: MEDICAL CERTIFICATION
I. Disease or condition directly leading to death (a) Myocardial Infarction
Antecedent causes:
Due to (b)
Due to (c)
4201

DEPARTMENT OF COMMERCE
Bureau of the Census

STANDARD CERTIFICATE OF DEATH [redacted]

Registration Dist. No. 2306 Division of Vital Statistics—State Board of Health State File No.
Registrar's No. 61 State of South Carolina

WRITE PLAINLY, WITH UNFADING INK—THIS IS A PERMANENT RECORD. Every item of information should be carefully supplied. AGE should be stated EXACTLY. PHYSICIANS should state CAUSE OF DEATH in plain terms, so that it may be properly classified. Exact statement of OCCUPATION is very important.

1. PLACE OF DEATH:
(a) County Greenwood
(b) City or town Greenwood (If outside city or town limits, write RURAL)
(c) Name of hospital or institution:
(If not in hospital or institution, write street number or location)
(d) Length of stay: In hospital or institution (Specify whether years, months or days)
In this community 61 Yrs.

2. USUAL RESIDENCE OF DECEASED:
(a) State SC (b) County Gwood
(c) City or town Greenwood (If outside city or town limits, write RURAL)
(d) Street No. Route 2 Box 14 A (If rural, give location)
(e) If foreign born, how long in U. S. A.? years.

3. (a) FULL NAME Willie Ann Patterson
3. (b) If veteran, name war
3. (c) Social Security No.
4. Sex Female
5. Color or race C
6. (a) Single, widowed, married, divorced Widow
6. (b) Name of husband or wife Robert Patterson
6. (c) Age of husband or wife if alive years
7. Birth date of deceased (Month) (Day) 1886 (Year)
8. AGE: Years 61 Months Days If less than one day hr. min.
9. Birthplace Greenwood SC (City, town, or county) (State or foreign country)
10. Usual occupation Domestic
11. Industry or business Self
Father: 12. Name Sam Holloway
13. Birthplace SC (City, town, or county) (State or foreign country)
Mother: 14. Maiden name Martha [illegible]
15. Birthplace SC (City, town, or county) (State or foreign country)
16. (a) Informant's own signature Willie Mae Holloway Patterson
(b) Address Route 2 Box 14 A Gwood
17. (a) Burial (Burial, cremation, or removal) (b) Date thereof 10-14-47 (Month) (Day) (Year)
(c) Place; burial or cremation Beulah Cemet.
18. (a) Signature of funeral director Robinson & Son
(b) Address Gwood SC
19. (a) 12-10-47 (Date received local registrar) (b) Mrs A. M. Greene (Registrar's signature)

MEDICAL CERTIFICATION

20. Date of death: Month 10 day 12 year 1947 hour 9 minute 30 AM.
21. I hereby certify that I attended the deceased from 10 11, 1947 to 10 12, 1947 that I last saw her alive on 10 12, 1947 and that death occurred on the date and hour stated above.

	DURATION
Immediate cause of death Coronary Occlusion	
Due to	24 hr.
Due to	
Other conditions (Include pregnancy within three months of death)	PHYSICIAN
Major findings: Of operations No	Underline the cause to which death should be charged statistically.
Of autopsy No	

22. If death was due to external causes, fill in the following:
(a) Accident, suicide, or homicide (specify)
(b) Date of occurrence
(c) Where did injury occur? (City or town) (County) (State)
(d) Did injury occur in or about home, on farm, in industrial place, in public place? (Specify type of place)
While at work? (e) Means of injury
23. Signature [illegible] (M.D. or other)
Address Greenwood SC Date signed 10-18-47

(10-726)

There are no Public Records kept of Birth in this State before 1915. This form when properly filled out and filed here, the certificate can be issued upon receipt of the fee. For Instruction See Reverse Side of This Form.

STATE OF SOUTH CAROLINA

APPLICATION FOR CERTIFICATE OF BIRTH

(With Instructions)

Full Name Will Patterson — Date of Birth Feb-25-1888

Male yes — Female — Where Born:

Race Colored — Nationality American — Near Epworth SC

Present Address Greenwood SC — County Abbeville

State SC

Full Name of Father Reuben Patterson — Age If Living — Age at Death 52

Nationality American — Where Born: Near Epworth SC

Race Colored — County Abbeville

Present Address Deceased — State SC

Maiden Name of Mother Katherine Holloway — Age if Living — Age at Death 70

Nationality American — Where Born: Near Epworth SC

Race Colored — County Abbeville

Present Address Deceased — State SC

A. If family Bible is used as proof

PHYSICIAN'S CERTIFICATE:

B. I hereby certify that I attended the birth of the above named child.

My Present Age.

Attending Physician.

(C, D, E, F, G, proof or other record.)

STATE OF

County of

Personally appeared before me

Who, being first duly sworn, deposes and says that

and that the statement of birth given above is true as copied from said record.

Sworn to and subscribed before me this

day of , 195

Notary Public.

STATE OF South Carolina — friend of family

County of Greenwood

Personally appeared before me Silas Brown (96 1/2)

who, being first duly sworn, deposes and says that he is 78 years old; that deponent is definite and positive of the above facts as to the date and place of birth, and that the statement of birth so given above is true.

Sworn to and subscribed before me this 21

day of May, 1953

Notary Public. — his X Silas X Brown mark

STATE OF South Carolina

County of Greenwood

Personally appeared before me Will Patterson

who, being first duly sworn, deposes and says that he is 65 years old; that deponent is definite and positive of the above facts as to the date and place of birth, and that the statement of birth so given above is true.

Sworn to and subscribed before me this 28th

day of May, 1953

Notary Public. — (Mrs.) Mary Payne Kinard

(OVER)

1. PLACE OF DEATH

County of Greenwood

Township of

or

Inc. Town of

or

City of Greenwood

Greenwood Hospital (Ward)

(If death occurred in a Hospital or Institution, give its NAME instead of street and number)

CERTIFICATE OF DEATH

STATE OF SOUTH CAROLINA.

Bureau of Vital Statistics

State Board of Health

Registration District No. 23a

Registered No. 163

(For use of Local Registrar)

File No.—For State Registrar Only. 29937

2. FULL NAME Nelson Patterson

Residence In City ... Yrs. ... Mos. ... Days.

PERSONAL AND STATISTICAL PARTICULARS

3 SEX Male

4 COLOR OR RACE Black

5 SINGLE, MARRIED, WIDOWED, OR DIVORCED (Write the word) married

6 DATE OF BIRTH (Month) (Day) , 1 (Year)

7 AGE yrs. mos. dys. If LESS than 1 day hrs. or min.

8 OCCUPATION (a) Trade, profession or particular kind of work Day Laborer

(b) General nature of Industry, business, or establishment in which employed (or employer)

9 BIRTHPLACE (State or Country) S C

PARENTS

10 NAME OF FATHER

11 BIRTHPLACE OF FATHER (State or Country)

12 MAIDEN NAME OF MOTHER

13 BIRTHPLACE OF MOTHER (State or Country)

14 THE ABOVE IS TRUE TO THE BEST OF MY KNOWLEDGE

Informant Mary Emery

Address [illegible]

15

Filed Jan 22 1919 WAWilliams LOCAL REGISTRAR.

MEDICAL CERTIFICATE OF DEATH

16 DATE OF DEATH Dec 15, 1918 (Month) (Day) (Year)

17 I HEREBY CERTIFY, That I attended deceased from Dec 5 1918, to Dec 15 1918, that I last saw h alive on Dec 15, 1918 and that death occurred, on the date stated above, at 4:00 P m. The CAUSE OF DEATH* was as follows:

[illegible]

Duration yrs. mos. 10 days

Contributory (SECONDARY)

Duration yrs. mos. days

18 Where was disease contracted if not at place of death?

Did an operation precede death? Date of

Was there an autopsy?

What test confirmed diagnosis?

(Signed) R. B. [illegible] M.D.

Dec 17, 1918 Address Greenwood S.C.

*State the Disease Causing Death, or, in deaths from Violent Causes state (1) Means of Injury; and (2) whether Accidental, Suicidal or Homicidal.

19 Place of Burial or Removal Mt Moriah

DATE OF BURIAL Dec 17 1918

20 UNDERTAKER W H Wright

ADDRESS Greenwood S.C.

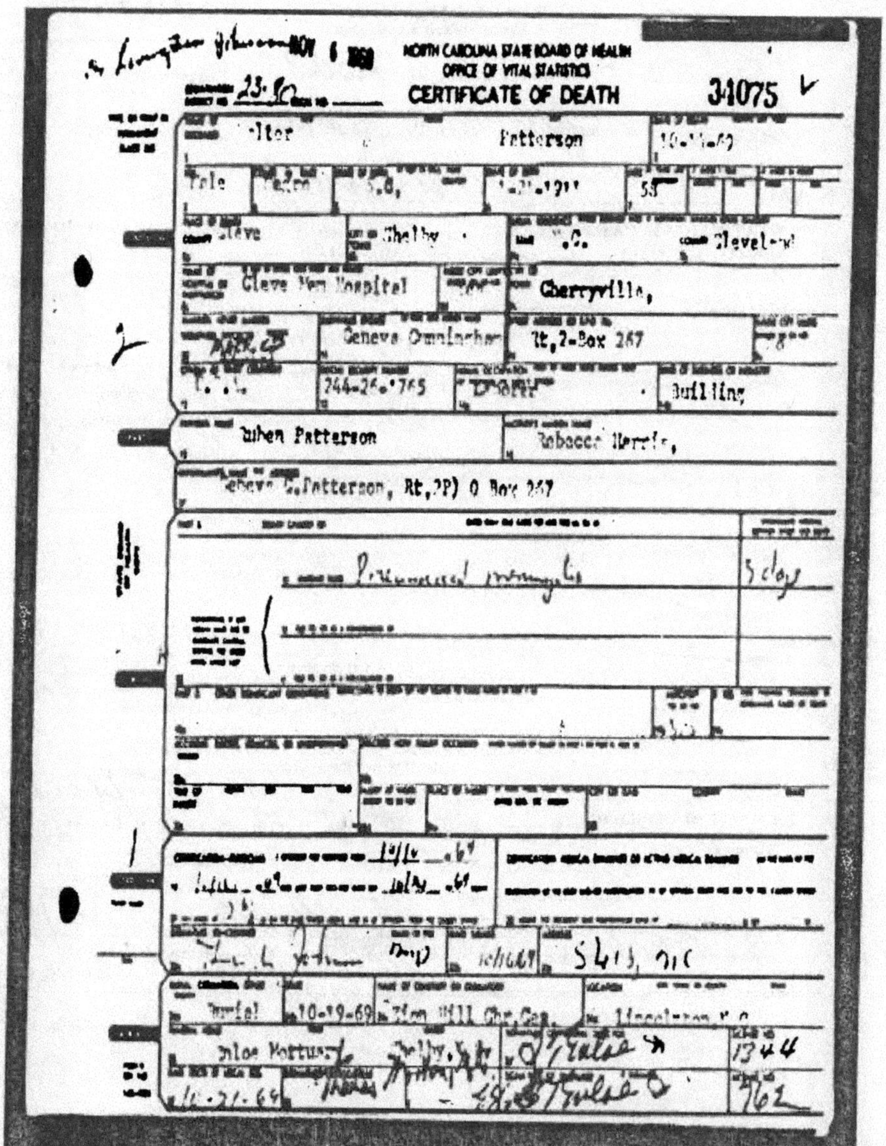

NORTH CAROLINA STATE BOARD OF HEALTH
OFFICE OF VITAL STATISTICS
CERTIFICATE OF DEATH

34075

Patterson

Cleve Mem Hospital

Cherryville,

Geneva Cunningham

Rt, 2-Box 267

Building

Patterson

Patterson, Rt, 2P) O Box 267

Consumption

Health Department of the City of Charleston.

"All Permits for the removal of the body of any deceased person from the City of Charleston for Interment, and all Burial Permits, and Permits for the Disinterment of the remains of deceased persons in the City of Charleston, shall be granted and signed by the Registrar."

Color

No Permit for Burial can be Obtained Without a Proper Certificate.

Male

All Physicians practising in Charleston (including those in public institutions) are requested to register their names in the Bureau of Vital Statistics.

Adult

CERTIFICATE OF DEATH. 961

1. *Full Name of Deceased,* (Write legibly and spell correctly. If an infant not named, give parents' name) Guy Patterson
2. *Age,* 34 *years,* — *months,* — *days.* *Color,* Colored
3. ~~*Single*~~ ~~*Married*~~, ~~*Widow*~~ *or Widower,* (Cross out the words not required in this line) 4. *Occupation,*
5. *Birthplace,* (State or Country) S o Ca (How long in the United States, if of foreign birth) 34 yrs
6. *How long resident in this City* 16 yrs
7. *Father's Birthplace,* (State or Country) S o Ca
8. *Mother's Birthplace,* (State or Country) S o Ca
9. **Place of Death,** (If an institution, please state the name.) *No.* 6 Philadelphia St *Street,* 5th *Ward.*
10. **I Hereby Certify** *that I attended deceased from* March 23rd 1881 *to* June 2nd 1881, *that I last saw* him *alive on the* 2nd *day of* June 1881, *that* he *died on the* 2nd *day of* June 1881, *about* 8 *o'clock,* ~~*A. M.*~~ *or P. M., and that* **the Cause of** his *death was:*

First (Primary), Phthisis Pulmonalis

Second (Immediate),

Time from Attack till Death: Write opposite each cause, if unknown, it should be so stated.

☞ All the above information should be furnished by the Physician.

Signed by Manning Simons [illegible] *Medical Attendant*

Address,

Place of Burial, Circular

Date of Burial,

Undertaker, [illegible]

Place of Business, [illegible]

Form No. 10

CERTIFICATE OF DEATH
STATE OF SOUTH CAROLINA.
Bureau of Vital Statistics
State Board of Health

File No.—For State Registrar Only.
16979

1. PLACE OF DEATH
County of Mc Cormick
Township of Bordeaux
or
Inc. Town of ....
or
City of ....

Registration District No. 4-5 PD
Registered No. .... (For use of Local Registrar)
(No. .... St.; .... Ward) (If death occurred in a Hospital or Institution give its NAME instead of street and number.)

2. FULL NAME George Patterson
Residence In City.....Yrs.....Mos.....Days.

PERSONAL AND STATISTICAL PARTICULARS	MEDICAL CERTIFICATE OF DEATH
3 SEX: male · 4 COLOR OR RACE: Black · 5 SINGLE, MARRIED, WIDOWED, OR DIVORCED (Write the word)	16 DATE OF DEATH: Sept 21, 19 26 (Month) (Day) (Year)
6 DATE OF BIRTH: .... 1...... (Month) (Day) (Year)	17 I HEREBY CERTIFY, That I attended deceased from July 20 1926 to Sept 21, 1926 that I last saw h ... alive on Sept 12 1926 and that death occurred, on the date stated above, at 4 P m. The CAUSE OF DEATH* was as follows:
7 AGE: 80 yrs. ..... mos. ..... dys. If LESS than 1 day ..... hrs. or ..... min.?	Brights Disease
8 OCCUPATION (a) Trade, profession, or particular kind of work: Farmer (b) General nature of industry, business, or establishment in which employed (or employer)	(Duration) ..... yrs. ..... mos. ..... days
9 BIRTHPLACE (State or Country)	Contributory (SECONDARY) ..... (Duration) ..... yrs. ..... mos. ..... days
PARENTS: 10 NAME OF FATHER: Jas Patterson	18 Where was disease contracted if not at place of death?
11 BIRTHPLACE OF FATHER (State or Country)	Did an operation precede death? ..... Date of .....
12 MAIDEN NAME OF MOTHER: Mariah Patterson	Was there an autopsy?
13 BIRTHPLACE OF MOTHER (State or Country)	What test confirmed diagnosis?
14 THE ABOVE IS TRUE TO THE BEST OF MY KNOWLEDGE (Informant) ..... (Address)	(Signed) M W Cheatham M. D. ..... 19... (Address)
	*State the Disease Causing Death, or, in deaths from Violent Causes state (1) Means of Injury; and (2) whether Accidental, Suicidal or Homicidal.
15 Filed ..... 19 ..... M W Cheatham LOCAL REGISTRAR.	19 Place of Burial or Removal · DATE OF BURIAL ..... 19...
	20 UNDERTAKER · ADDRESS

(Margin:) N. B.—Every item of information should be carefully supplied. AGE should be stated EXACTLY. PHYSICIANS should state CAUSE OF DEATH in plain terms so that it may be properly classified. Exact statement of OCCUPATION is very important. See instructions on back of certificate.

Joe Patterson

While seeking an interview with an ex-slave today, the writer was directed to a certain house where an old man lived. Entering the premises by the rear, he observed an old man helping a woman who was washing some clothes. He was stepping around quite lively, carrying water and emptying one pot after another of the dirty water already used by the woman. After he had sufficient water for his wife's needs, he asked the writer to go with him to the front porch where he could be quiet and talk. He stated that he was large enough during the Civil War to wait on the soldiers when they would come to his master's home for something to eat, which was at Kilgore's Bridge on Enoree River, and that his job during The slavery days was to wait on the white folks and watch the plantation.

He also stated that his father was a full-blooded Indian who was sold to his master by Joe Crews, the biggest slave trader in the country. His father was stolen somewhere in Mississippi, along with other Indians, and sold into slavery with the "niggers." He said his father told him he was stolen by Joe Crews when he was a young buck. At that time, his father went by the name of "Pink Crews," but after he was purchased by Mr. Joe Patterson, his name became "Pink Patterson." He stated that his mother was a white woman who came from Ireland and was working on the Patterson farm. She was not a slave, but was married to his father by his "Marster."

They lived in a one-story, one-room log cabin which had a dirt floor. The whole family of 18 children and parents lived in this small house. They were comfortable, however, and all had good health. He stated that he had not been sick for fifty years, and that the only trouble with him now was a broken foot, the result of a railroad wreck about forty years ago. He said his foot still gave him trouble in bad weather.

He said he had not been conjured at all, but had just gotten his foot broken. "Conjuring and ghosts are all foolishness anyhow." The nearest he ever came to seeing a ghost was one night when he observed a "white thing moving back and forth across the branch." He had with him his brother's cap and ball pistol, and he shot at the object two or three times, knowing that his dogs would come to him if they heard the shots. Two or three dogs came up and recognized him. He told one bull dog to go to the white think and see what it was. After the dog had been all around the place where the thing was moving, he knew there was nothing there to frighten him. Next morning, he went out to see the object and found it to be a small tree with white leaves waving in the breeze.

"Going back to slavery times, he said that on most plantations were kept squirrel dogs, 'possum dogs, snake dogs, rabbit dogs and "nigger" dogs. Each dog was trained for a certain kind of tracking. He used to train the "nigger" dogs which were used to track slaves who had run away from the plantation. He said he had two dogs that were sure never to lose the scent when they had taken it up. "If I put them on your track here and you went to Greenville, they would track you right to Greenville."

He said his "Marster" never chained his slaves but he had seen slaves on other plantations wearing chains to keep them from running away.

"People don't work like they used to, and this thing of higher education is ruining niggers. All their learning teaches them is how to beat a man out of a dollar and how to get out of work. It teaches them to cuss, and it teaches these young girls how to make easy money. As old as I am, I've been approached by girls I didn't know and asked for a dollar. Now that thing won't do. I believe in teaching children how to read and write, but don't go any further than that. I've never seen a moving picture. Once a man offered to give me a ticket to a movie, but I told him to give me a plug of tobacco instead."

When asked if he thought colored preachers should be educated, he replied that when they are educated they learn how to steal everything a man has, if they can."

You remember reading about Joe Crews and Jim Young-what they did in this state? Well, they tried to lead all the niggers after the war was over. I was the one who got Jim Young away from the whites. I carried him to Greenville, but he got back somehow, and was killed. Joe Crews was killed, too. The Ku Klux was after them hot, but I carried Jim Young away from them. You know, the Yankees was after getting all the gold and money in the South. After the war, some Yankee soldiers would come along and sell anybody, niggers or whites, for a gun. They were trying to get on to where the white people kept their money. If they caught on, they would go there and steal it. You know, there wasn't any banks, so people had to keep their money and gold in somebody's safe on some big man's place. These men in selling guns was trying to find out where the money was hid."

When asked about hunting, he said that hunting in slavery days was not like it is now, for a man could hunt on his own place then and get plenty of game. There were plenty of wild hogs in those days, as well as wild turkeys, rabbits and squirrels. Some of the hogs were so wild that no one dared to go into a pack of them, for they had tusks six inches long, and could tear a man to pieces. A man could shoot a wild hog and have no trouble over it. Cattle, he said, ran wild and were dangerous at all times."

When you buy something now, you haven't got much. I bought a cake of soap for my wife but it was a small thing. When we used to make our own soap on the plantation, we had plenty of good soap."

He said his father followed his master and others to the war, and he drove artillery wagons at times. At Appomattox, his father told him that he drove wagons over dead soldiers piled in ditches. His father lived to be 111 years old. After he and his father were set free, they remained with Mr. Joe Patterson to help him make that year's crops; then they moved to another place.

He heard that work was plentiful in Spartanburg, and he moved here and did various kinds of work. He said that he was not as strong as he used to be, but that he could still do a full day's work except when his foot troubled him.Uncle George was quite polite and seemed glad to talk of old times. He observed, though, that in old times people would speak to him. "You go up to a crowd now, and they won't speak. They won't notice you."SOURCE: George Patterson, 653 Peachtree Street, Spartanburg, SC. F.S. DuPre. From *Born in Slavery: Slave Narratives from the Federal Writer's Project, 1936-1938.* Project 1835-(1), FOLKLORE, Spartanburg, SC, District No.4, May 27, 1937. Edited by R. V. Williams.

Who Killed Joe Crews?

The Reconstruction of South Carolina was a difficult period in the history our state. A great deal of pain, agony, and the destruction of a considerable amount of property took place. It is not wise to judge history, for there are always victims and I suspect that almost always we are all victims, some simply make a vocation of applying blame and others of accepting guilt, history simply is, man is the revisionist. Most of the death that took place was not the result of vigilante action by either side but was usually the result of armed confrontation between two well armed groups. Discounting these major events which include Hamburg, the incident at Cainhoy, the Charleston riots, and the Ellenton tragedy in the 1876 election, we are left with a very few incidents of what could be called murder. There were a very few men actually killed on either side, in cases of ambush or lynching by persons unknown and the victims appear to be about equally divided in party affiliation. Of this handful, only a few remain unsolved. The Freedman's Bureau reported about 100 incidents between 1865-1870 that involved violence. Most of these represented disputes of a personal nature, and framed within the context of that time are typical of the disputes people always have with one another. In many cases local magistrates simply refused to act on these complaints and those who were republicans turned to the Union forces for action. There are stabbings and shootings and fights over property and a handful of murders by parties unknown. For more information see the Records of the Assistant Commissioner for the State of South Carolina, Bureau of Refugees, Freedmen and Abandoned Lands, 1865-1870. National Archives Microfilm Publication M869 Roll 34. In one of the best known of these, motive still eludes us, as does the assailant, Although the latter maybe more widely known today than ever before. This event occurred in Laurens County and had an impact on Greenville County, particularly Lower Greenville County, as events like these rarely know geographic borders. In that death the victim was Joe Crews. To evoke an old southern saying, many people thought that Joe Crews probably needed killing anyway, and here is his story. However it is unique how even today, the death of Joe Crews continues to play itself out.

Prior to the War Between the States, Joe Crews had made an excellent living trading in slaves. As a general rule, he purchased as far north as Virginia and sold in South Carolina. At least one account of his having stolen slaves does survive but this may very well be information developed after the fact, to support a particular point of view. For you see, Joe Crews was the consummate scalawag. Following the war, Joe Crews involved himself in Reconstruction politics in Laurens County. As a radical, he had few equals in the state and was considered one of the best thieves, of that era of thieves, called reconstruction in our state. Crews was a member of the Union League or the Loyal League, as it was known in some places. The Union League was the Secret Society of the Republican Party, not unlike its Democratic counterpart, the Klan. A.A. Taylor, in his book, The Negro in South Carolina calls attention to Joe Crews on page 190 of his work in this way, "Unfortunately, too, the Negro militia was not wisely handled. Under Joe Crews, the white Lieutenant Colonel, the militiamen did so much parading and unnecessary talking about, how they intended to keep order, and they made so many arrests that whites organized in opposition." Even the revisionists give Joe Crews credit for being able to stuff a ballot box, as he did so well in the 1870 election. That event brought about the riot in Laurens that many thought to be his undoing.

"Crews met a violent end. He was assassinated from ambush, in 1875, while riding on a public highway. His slayer was never apprehended and hence it is not known whether the killing was due to his political activities or to someone's desire for personal revenge. Crews, for reasons besides political, had made many enemies." Ousting the Carpetbagger from South Carolina Williams is a bit more romantic, "Crews leaving his home early one morning in his buggy, was waylaid and shot with six buckshot. He was found alive and it was said knew who fired the bullets into him. One of the many gruesome mysteries of that time was the story that the killer visited the man he had killed, was welcomed with a cordial handshake and assured that his act would not be revealed and Crews died with lips sealed. That is one of several mysteries of the Reconstruction period that never will be explained until the last judgment. General opinion fastened the crime - if it was a crime - on one man but nobody dared name him. It is a curious psychology that, after fifty years and with all the parties immediately concerned dead and buried long ago, I will not even now tell to anybody the name of the man who was generally suspected, so deeply was the need for caution and secrecy ground into us. Years later he rose to prominence and importance but his name does not appear in any record connected with the killing of Crews."

Crew's activity in the legislature had also done nothing to endear him to the people he served. It is widely stated that he not only accepted bribes he was often involved in the encouragement of others to violence, like so many others of that time, he was the among the first to say that matches were plentiful and cheap and fires easy to start. However, by far his most notorious activity was his denouncement of about forty men that he characterized as Klansmen following the riot of 1870 in Laurens. Laurens was reported to have little or no Klan activity, so the arrests in 1872, some two years after the event in question, while not surprising, did open the door for more violence to follow. (Ousting the Carpetbagger) Of the 40 or so men arrested none were convicted but many were held for a rather long time. The men were held in both Charleston and Columbia for the death of one and the injury of two others in that 1870 event. All the men who had been arrested were pardoned in 1873 by President Grant but not without exacerbating a difficult situation, especially in Laurens County.

Given this history Crews ambush in 1876 came as a surprise to no one. What was a surprise was that he clearly knew who had killed him and refused to identify the man or men involved in the days he lingered prior to his death.

I knew of Crews and his murder from the accounts of A.B. Williams in Hampton and his Redshirts and the accounts in Ousting the Carpetbagger from South Carolina. I also knew the death was written of as the fault of parties unknown. Not long ago, a close friend announced to me that he knew who killed Joe Crews. Somewhat taken back I listened as he said that Crews was an ancestor of his, as was the man who killed him. He explained, as is the custom in the south, this was a closely guarded family secret and it was passed to him mouth to ear by his father. His father had taken him to a place near the home of Crews, told the story and shown him the places. Even today we guard our own with both our heart and our head in the south. He identified the man responsible as Wash Shell, a relative and pointed me to a copy of A.B. Williams book that he donated to a library with his handwritten footnote concerning the event.. "Note: Wash Shell- lived on East Main, next to Celene Bryson's home. Wash Shell killed 'Uncle Joe' Crews near by-pass across from Forest Lawn (Cemetery in Laurens) Crew's home was at end of road near/at Badger's Trestle."

When reading, Hurrah for Hampton, a new book, by Drago I found in one of the WPA accounts a mention of the death of Joe Crews. It was with even more surprise that I discovered one of these accounts that identified the dead man as Jim Crews and the responsible party as Wash Hill. Drago observed in his book, "George Fleming ofLaurens County had nothing but disgust for white Republican Joe Crews, a former slave trader and other scalawags like him. For this young Democrat, such men 'had done made all de money dey could off selling niggers, so dey thought dey could make some more by making agreements wid de Republicans.'" George Fleming goes on to say this concerning Crews, Old Jim Crews was killed to, at that time. Wash Hill was the one that got him. He was shot at Crew's Branch, it wasn't long after that till things began to settle down, for the Democrats shore did lick the Republicans." I suspect that even as late as 1937 there was reluctance by the person recording the event to name the correct name or that the dialect was misunderstood, which seems unlikely. For whatever reason, it appears that one hundred and fifty years later we still have some idea of who shot Joe Crews and that it was a family member and possibly a family matter. As Faulkner said so well, In the south, the past is never dead, it isn't even past.

Enoree River

Kilgore's Bridge on F.noree River

SLAVERY IN NEWBERRY COUNTY, SOUTH CAROLINA

"I was born in the town of Newberry, S.C. *I do not remember slavery time, but I have heard my father and mother talk about it. They were Washington and Polly Holloway, and belonged to Judge J.B. O'Neall. They lived about* 3 *miles west of town, near Bush River. An old colored man lived nearby. His name was Harry O'Neall, and everybody said he was a miser and saved up his money and buried it near the O'Neall spring. Somebody dug around there but never found any money. There were two springs, one was called 'horse spring', but the one where the money was supposed to be buried had a big tree by it.*

"I married Sam Veals, in 'gravel town' of Newberry. I had a brother, Riley, and some sisters.

"We would eat fish, rabbits, 'possums and squirrels which folks caught or killed. We used to travel most by foot, going sometimes ten miles to any place. We walked to school, three or four miles, every day when I was teaching school after the war. I was taught mostly at home, by Miss Sallie O'Neall, a daughter of Judge JB. O'Neall.

"My father and mother used to go to the white folks' church, in slavery time. After the war colored churches started. The first one in our section was Brush Harbor. Simon Miller was a fine colored preacher who preached in Brush Harbor on Vandalusah Spring Hill. Isaac Cook was a good preacher. We used to sing, 'Gimme dat good ole-time religion'; 'I'm going to serve God until] die' and 'I am glad salvation is free'.

Saturday afternoons we had 'off and could work for ourselves. At marriages, we had frolics and big dinners. Some of the games were: rope jumping; hide and seek, and, ring around the roses. Of course, there were more games.

Some of the old folks used to see ghosts, but I never did see any. "Cures were made with herbs such as, peach tree leaves, boiled as a tea and drunk for fevers. Rabbit tobacco (life everlasting) was used for colds. Small boys would chew and smoke it, as did some of the old folks.

"I have seven children, all grown; fourteen grand-children, and several great-grand-children.

"Judge O'Neall was one of the best men and best masters in the country that I knew of. I think Abraham Lincoln was a good man, according to what I have heard about him. Jeff Davis was the same. Booker Washington was a great man to his country and served the colored race.

"I joined the church because I believe the bible is true, and according to what it says, the righteous are the only people God is pleased with. Without holiness no man shall see God. "

Source: Mary Veals (72), *Newberry,* S.C. *Interviewed by:* G.L. *Summer, Newberry,* S.C. *May 20,1937.*

(Slave Narratives: a Folk History of Slavery by Work Projects Administration, February 24, 2009: EBook 28170)

Newberry History

Newberry is a community filled to its borders with history: ancient Indian sites, battlefields of the American Revolution, historic plantations, and beautiful homes. European settlers (primarily German, Scotch-Irish, and English) began appearing in great numbers in the 1750's. Newberry County, formed from the Ninety-Six District in 1785, was once described as the largest tract of unbroken farm land in South Carolina. The origin of the county's name is still unknown. It is likely an alternate spelling for the English town "Newbury," but the popular notion has always been that the surrounding fields and forests were as pretty as a "new berry." Although cotton was the primary crop before the Civil War, today's farmers rotate crops such as com, millet, wheat, and soybeans. In addition Newberry has dairy, poultry, and cattle farms, as well as many acres of controlled reforestation.

The town of Newberry was founded in 1789 as the county seat. Its site was chosen because of its nearness to the center of the county. By the coming ofthe railroad in 1851, Newberry had become a thriving trade center. Lutheran-supported Newberry College was established in 1856 and has been an important part of the community ever since. Although the Civil War interrupted the growth of the town and dramatically changed its social order, a stronger community emerged which continued to thrive. Industry, in the form of cotton mills, was introduced to the town in 1881. Although the face of the town has changed because of fires, storms, and former economic slumps, the City of Newberry today retains diverse historic buildings and a revitalized downtown.

Since rivers form the boundaries of the county, other communities developed at highway crossroads and, later railroad depots. Among the towns incorporated as a result of the Greenville and Columbia Railroad were Peak, Pomaria, Frog Level (now Prosperity), Silverstreet and Chappells. A branch railroad to Laurens in 1854 had depots at Jalapa and Kinards. In 1890, the arrival ofthe Columbia, Newberry and Laurens Railroad prompted the incorporation of Little Mountain. Whitmire, a trading center on the Enoree River, was incorporated in 1891 when the Georgia, Carolina and Northern Railroad came through. Aside from the City of Newberry, Prosperity and Whitmire are the most populous towns in the county.

Many interesting and colorful personalities have made a mark on Newberry's history. Emily Geiger, a young woman living in what is now eastern Newberry County, rode her way into the history books when she delivered a message from General Nathaniel Greene to General Thomas Sumter during the American Revolution. Tales also abound about a Quaker girl named Hannah Gaunt who helped defend her father's house against a Tory attack. John Belton O'Neall was a prominent judge in Newberry until his death in 1863. Among his many accomplishments is The Annals ofNewberry, an early history of the county. Job Johnstone (1793-1862), a Newberry lawyer, served as Chancellor in South Carolina for thirty-two years and later served on the State Court of Appeals. Another Newberry lawyer, John Fletcher Hobbs, left for Australia in 1882 and, by 1893, had become chiefoftwo tribes of cannibals. Marie Boozer gained notoriety for her great beauty, and her exploits (after leaving Newberry) were the inspiration for two books: La Belle and Another Jezebel. Coleman L. Blease (1868-1942) was the only permanent resident ofNewberry to be Governor of South Carolina. A lawyer, Representative and United States Senator, he was elected Governor in 1910 and 1912. Interestingly, his two opponents in 1912 were also from Newberry.

Among the many scenic and historic sites in the county are: The Rock House (pre-Revolutionary, the oldest house in the county); Quaker Cemetery (used from the 1760's- 1820's); Tea Table Rock (site of a British encampment during the Revolutionary War); St. John's Lutheran Church, Pomaria (1808); Little Mountain (800 feet above sea level, highest point in county); Gauntt House, Newberry (1808, oldest home in city); Hardy House, Maybinton (1825, typical of early nineteenth century); Pomaria Plantation, Pomaria (1826, site of a well-known nursery); Old Court House (1851); Newberry College (founded 1856); Jasper Hall, Whitmire (1857, fine antebellum residence); Rosemont Cemetery (1863); Newberry Opera House (1881); Oakhurst, Newberry (1891, a fine Victorian home); Lake Murray; and Lynches Woods (a scenic road winds through the forest.

Annals of Newberry

IN TWO PARTS

PART FIRST

JOHN BELTON O'NEALL, LL.D

PART SECOND

JOHN A. CHAPMAN, A.M.

COMPLETE IN ONE VOLUME

CLEAFIELD
COMPANY

The folllowing is the entire chapter 26 of John Belton O'Neall's The Annals of Newberry, the chapter concerned with Judge O'Nealls father, Hugh O'Neall, and his ancestors. This chapter is the principal source of information on Hugh O'Neall, the immigrant, his wife, Anne Cox, and their children.

Errors of spelling and punctuation have been retained as in the original, which was published in 1892 by Aull and Houseal, Newberry, South Carolina. A reprint was published by the Genealogical Publishing Company,Baltimore, in 1974.

To-day I propose to sketch, imperfectly, I know, and perhaps *too partially,* the life and times of Hugh O'Neall, one of Newberry's oldest and best citizens.

He was born on Mudlick, Laurens district, at the place, late the property of John Armstrong, deceased, on the lOth of June, 1767. He was the second son of William O'Neall and his wife, Mary Frost. They removed after the birth of their two first children, Abijah and Sarah, to South Carolina. The family remained in Laurens until after the birth of Henry, the third son, who, I think, was born in '77; indeed I think they did not remove to Newberry until 1779. The family consisted of six sons; Abijah, Hugh, William, John, Henry and Thomas, and one daughter, Sarah, all of whom lived to rear families. Abijah removed in '99 to Ohio, near Waynesville, Warren county. Sarah married Elisha Ford, and removed to Shelby county, Kentucky. William died on Bush river; his body rests in the graveyard of Friends, near Mendenhall's Mills. John, Henry and Thomas removed to Indiana. They have all been gatherered *[sic]* to their rest, leaving families more or less numerous.

William O'Neall's father's name was Hugh; he was, I think, a midshipman in, or at any rate he belonged to, the English navy, and not liking his berth, while at anchor in the Delaware he jumped overboard, swam ashore, and landed near Wilmington, as well as I remember, at the little Swedish town of Christiana; this took place about 1730; here he lived many years, and married Annie Cox. On landing, to escape detection, he altered the spelling of his name, either from O'Neill or O'Neale to O'Neall; the latter is the tradition. His family consisted of William, James, Hugh, Henry, John, Thomas, a daughter, Mary, and a posthumous son, George. In his life time he removed to the Susquehanna, and there he died; his family thence removed to Winchester, Virginia; there William married his wife, Mary Frost; and there, as already mentioned, his two eldest children were born.

The family, with the exception of James and George, removed about 1766 to South Carolina. Thomas died at Parkins (now Crofts) on Saluda, and was the first person buried in that graveyard. Hugh married a Parkins, and settled and died at Milton, Laurens district. Henry married a Chambers, lived in Laurens, and there remained till the close of the revolution, when he removed to Florida, and settled the place at the mouth of St. Mary's river (where his grandson, the Hon. James T. O'Neill, now resides); he (Henry O'Neall) was killed in an attempt to seize an outlaw soon after his removal; he left a large family-James, Eber, Thomas, William, Henry, Asa, Hugh and Margaret; all are dead except Margaret, now Mrs. King, of Georgia; none had families except Eber, William and Margaret.

William O'Neall was a Friend; when he joined that body of religionists is not known; his wife also belonged to the same; his brother, Hugh, inclined the same way; so did his wife and the entire Parkins family. In the revolution neither of these brothers took any part, except to bury the dead, heal the wounded, and do good wherever they could. James and George belonged to the American army; the former was a Major in the Virginia line, the latter a common soldier. Both served the entire war, and at its close, ignorantly supposing that the O' in their names was some aristocratic distinction, instead of meaning, as it really does, the "son of," struck it off and wrote their names Neall. James settled at, or near, Wheeling, Virginia; George in Jessamine county, near Nicholassville, Kentucky: they both have been dead many years; each left families surviving them. I should be proud if their descendants would resume the O', which rightfully belongs to their name.

Henry and John, unfortunately, sided with the tories. Henry, it is said, after his determination was made, and he had accepted a Major's commission in the British army, passed into Virginia to see his brother James, and proposed, if they should ever meet in battle, that they would treat each other as brothers; but the stem republican would accept no such amnesty; "in peace, brethren; in war, enemies," was his reply. Fortunately, they never met in arms.

John married Grace Frost, the sister of his brother William's wife; he was a captain in the Tory forces, and was killed in a skirmish with Colonel Roebuck, in Union district; he left two daughters, Sarah, and, I think, Rebecca; his widow married a well known citizen of Pendleton, Mr. Crosby. Mary married Frederick Jones. She had an only son, Marmaduke, who will be remembered as a resident of Laurens district, in the neighborhood of Milton.

Having thus stated his ancestral families, and his father's, I now propose to give a sketch of the life and times of Hugh O'Neall.

He went early to school, he learned rapidly: most of that which he learned was with a Virginian, Benj. Smith. In his school, in company with Major John Griffin, James C. Griffin, the Williams', Cress wells, Caldwells, he acquired the common elementary education, reading, ,writing and arithmetic. Reading, all his life, was his great delight; he began early and continued late. His memory was early developed and long retained; often in middle life, and even in old age, has he recited many passages in the tragedy called the Battle of the Boyne, which he had read when a boy among his uncle Henry's books. The poem called Sir James the Ross, was another read in the same way, which he often repeated. One of his early exercises was a riddle, propounded by his teacher, Mr. Smith, pretty much as follows, viz.:

Beneath the heaven, a creature once did dwell,
As sacred writers unto us do tell;
He lived, he breathed in this lower world, it is true,
But never sinned, nor any evil knew;
He never shall be raised from the dead,
Nor at the day of judgment show his head;

He never shall in heaven dwell,
Nor yet be doomed to feel the pains of hell-
yet in him, a soul there was, that must
Be lost, or live above, among the just.

This he solved by giving Jonah in the whale's belly, and often repeated it in manhood, and age. Benj. Smith was one of the Virginia troops on service in '76, perhaps against the Cherokee Indians, under Christie, or in Gen. Lee's projected invasion of Florida, and was either left, as unable to travel on the return march, or discharged. From the description given of him, he was both a man of talents and education. His Impress was to be seen on all his scholars.

In, I presume, the year '78 was the great May frost, which took place on the 4th, and utterly destroyed vegetation and the crops; a small crop of late wheat was saved by William O'Neall. In the same year was the total eclipse of the sun. The total darkness was so great that chickens went to roost. The upper part of South Carolina, as has been frequently and justly said, scarcely knew that there was war, until the siege of Charleston. The incursion of the Cherokees on the 30th of June, '76, drove the settlers nearest the frontiers from their homes. William O'Neall, with his family, fled from Mudlick to Benj. Pearson's, near Kelly's old store, now Springfield. Often has Hugh O'Neall pointed out the old field west of Dr. Wm. Harrington's attempted settlement, in

34

Frost's old field, as being then in cultivation, and stated the fact, that he had swam in Pearson's Mill pond on Scott's creek, where Fernandes' pond lately was.

In 1780, when Charleston fell, William O'Neall and family lived at the place, about a mile west ofBobo's Mills, and on the southwest side of Bush river. He then owned the mill, known for thirty years as O'Neall's, now owned by Dr. J. E. Bobo, about one and a half miles below Mendenhall's. Hugh O'Neall, the subject of this memoir, was then thirteen years old; yet his services were so necessary to his father, that he either attended entirely to the mill, or was a constant assistant. In that way, although no actor in the revolution, yet he became fully informed of most ofthe events ofthat dark and bloody period. The mill was the most public place in that section of the country. Across Bush river, at that place, was the most common thoroughfare from the Congaree and Charleston to pass south beyond Saluda, and west to Little river and Ninety-Six. There, were often halted the scouts, sometimes the armies; there, too, were provisions seized, as want, or power dictated. There, as he often afterwards said, did he learn to hate the proud, overbearing character of the British officers. There he heard narrated the accounts of the many deeds of violence and blood with which the country was overspread. The various sketches of men and events heretofore given are in a greater, or less degree, dependent upon his wonderful memory for their accuracy.

To give a true sketch of the bloody partisan war from 1780 to 1783, would be a most Herculean task; much of it has been already done in the different biographical sketches and anecdotes already published. Blood and plunder were the watchwords of many ofthe different parties who swept over old Ninety-Six. "Each party," (as Gen. Moultrie, in his Memoirs, vol. 2d, p. 301, appropriately says,) "oppressed the other as much as they possibly could, which raised their inveteracy to so great a height that they carried on the war with savage cruelty; although they had been friends, neighbors and brothers, they had no feelings for each other, and no principles of humanity left." At page 303 he says: "The conduct of these two parties, (whigs and tories,) was a disgrace to human nature, and it may, with safety, be said that they destroyed more property, shed more American blood, than the whole British army." The pictures thus given in a few words, are, unfortunately, too true, and ought to teach us to beware of the tendencies to civil war, which I sometimes fear are too much encouraged.

The march of the British army was marked by wasting and ruin. When Greene passed, with his ragged Americans, forbearance and pity for the people marked his course; plunder, cruelty and oppression, he sternly forbade. When a battalion of Tarlton's command, in his attempt to strike Morgan, as he supposed, in the neighborhood of Ninety-Six, (as is stated in a note to No.5,) encamped at William O'Neall's, everything was seized and treated as if it all belonged to them, the fences were burned to make camp fires, the cattle were butchered for beef, the officers billeted themselves on the unpretending Quaker family, without money and without price. When a part of Greene's army, on their retreat from Ninety-Six, passed the mill, everything needed was paid for, and perfect order prevailed.

The marauding scouts entered every dwelling, and carried off everything which suited them, bedding, clothes, provisions; often were families left without food or raiment; sometimes the houses were burned, and women and children turned out with no covering, save the forest and the heavens.

These scenes passed before the eyes of the youthful Quaker, Hugh O'Neall; his brave ancestral blood often boiled almost over at the wrongs and oppression which he witnessed, and to which he was called to submit. Yet the teachings ofhis parents, *peace, peace,* kept him quiet, and day after day he was seen at the mill, providing for his father's family and the neighborhood's necessities, as well as he could, until, at last, *peace, smiling peace,* and glorious liberty came to bless South Carolina with *law and order.*

Hugh O'Neall attended the mill, drove his father's wagon, or labored in the farm until his father's death in 1789. He and his elder brother Abijah were the executors of his father's will and upon them devolved the care of a large real estate, their mother and a family of young boys. The elder brother, Abijah, being married, much of the burden devolved on Hugh. For three years he devoted himself untiringly to the discharge of his duties. Many of his adventures in wagoning between Newberry and Charleston, and in Charleston, would, if I had time or space, be interesting. I may state two: He and his brother Abijah were in Charleston when the old State House, now the Court House, comer of Broad and Meeting streets, and all that section of Charleston was burned. They had one or more wagons, and were employed to haul goods from the burning district to places of safety. Having made several successful trips, as Hugh was returning, and about to pass again into the circle of fire, his leader's bridle was seized by a policeman on duty, and he was told: The houses near you will be instantly blown up! He turned his team, quick as thought, in the crowded streets, and was soon in the wagon yard and safety. Neither the persuasions of his brother nor the tempting wages could again tempt him into such peril.

Roads, bridges and ferries were then, not as they are now, (though now bad enough.) Mud holes, crazy bridges, streams in flood, and badly managed ferries had to be encountered. He and his brother-in-law, Ford, were on their return from Charleston, with separate teams. Ford was in front. He struck the Four Hole swamp, covered with water. When he reached the bridge it was floating; he thought he could, however, pass it, and with the bold, adventurous spirit of a backwoods man, well tried in the revolution, he made the attempt. The plank gave way under his horses, and into the stream they went. To cut them (except one) loose, and to swim them out was but a few minutes' work for him and his equally daring companion, Hugh. One horse, the old and favorite leader, was patiently lying across the sleepers of the bridge; to relieve him it was necessary to roll him over into the water. This was done by seizing his legs and literally turning him over. As he went, with one strong movement of his hind leg he threw Hugh twenty feet, into ten-feet water. This was, however, no serious matter, for he and the horse were soon on terra firma.

During this period, and for years after, tobacco rolling was a common mode of carrying tobacco from the upper country to Charleston. A tobacco hogshead was rimmed, so as to keep the bulge from the ground; a cross piece was made fast to each end; in them were inserted wooden gudgeons, which worked into a square frame, embracing within it the whole hogshead. To this were fixed single-trees and a tongue, and, thus prepared, the owner mounted on one of two horses geared to it, and leading the other, with his fodder and corn stowed between the frame and hogshead, moved on a free and independent roller to Charleston; and there leaving his hogshead, with his money for it, or a tobacco certificate, he returned, the sauciest mortal ever seen. Some rollers from Long Cane, Abbeville, and, therefore, called Long Canaans, met with an Edgefield man, (Clarke Spraggins,) and a companion, between Orangeburg and the Four Holes, attacked them first with words, and then were about to try blows. Numbers prevailed, and Spraggins, (though one of Butler's old soldiers,) and his companion had to fly. In his flight Spraggins sprang off his horse, picked up a lightwood knot, and knocked down senseless the foremost pursuer. The rest halted, and supposing their companion slain, desired to know who and whence was the slayer. Spraggins swore he was from "killman," and was going to "killmore."

In 1792 Hugh O'Neall married Anne Kelly, the third and youngest daughter of Samuel and Hannah Kelly, of Springfield, Newberry. He settled about a mile below the mill which, by his father's will, was devised to him. Subsequently he made an exchange with brother, William, and fixed his residence in about two hundred yards of the mill, on a hill northeast of the same, From 1792 to 1800 he attended to his own mill, and by untiring industry created the means to rebuild it and to lay up a sum sufficient to embark in the mercantile business with Capt. Daniel Parkins. During this period was the great Yazoo freshet, in January, 1796, which has never been equalled or surpassed, unless the disastrous freshet of August, 1852, did so. Often has Hugh O'Neall described that freshet to the writer. In two respects it resembled the freshet of August, 1852: it was a freshet upon a freshet, and, like the latter, it spread ruin everywhere. Mills, dams and bridges went before it. Compte's bridge across Broad river, three miles above Columbia, just finished in apparently the most secure way, went. It is said the owner, a Frenchman, was upon the bridge, looking at the raging torrent, and impiously exclaimed: "Aha, God Almighty does think we build bridges out of com-stalks." Scarcely were the words uttered, until the cracking timbers gave notice that its end was at hand. With difficulty the owner reached the land. Hampton's bridges across the Savannah at Augusta and Saluda, were swept away. Fortunately, O'Neall's mill, which was just rebuilt, with its dam, escaped uninjured. Would that some certain memorial of that flood had been preserved. We would then compare it with that of'52, and thus learn a lesson of wisdom.

During this same period, or possibly in '93, certainly before April, he and Mercer Babb visited the quarry of Georgia burr millstones, in Burke county. He did not contract for a pair, but Mercer Babb bought, and started in his mills, now Mendenhall's, the first pair of burr stones ever run in the district. They were there used for many years, and when Dr. Mendenhall, in '27, started his merchant mills at the same place, the old Georgia burrs were refitted and again started, to manufacture flour.

Hugh O'Neall always affirmed that, with a good pair of Cloud's creek stones, he could make as good, if not better, flour than could be made with the best pair of burr stones.

On this trip he and his friend encountered a flood in the Savannah and Saluda rivers, then considered a great freshet, but not to compare with the subsequent one of '96.

In 1800 Hugh O'Neall embarked in the mercantile business, as the partner of Daniel Parkins, and most successfully pursued it until the death of the latter, October, 1802. It may be well *here* to pause and look over the statistics of the country at that time, (if! can use such a word in reference to the means and commerce ofthat period.) Cotton, in 1800, was beginning to be cultivated for market. In 1801 Hugh O'Neall started a water cotton-gin, made by William Barret.

The plates for the saws were made at William Coate's shop. No machine ever ran with greater power or more success, although the first person, Joseph Wright, who attended to it had his hand tom all to pieces by the saws. Remittances were then made to Charleston in specie. Dollars were carefully packed in a box and put on board a wagon owned and driven by a careful, responsible man. The writer recollects aiding in counting, at Capt. Parkins', a large amount of silver, to be sent by Isaac Mills' wagon. Up to the year 1806 the upper country, and particularly Newberry, furnished flour, bacon, beef, cattle, butter, beeswax, skins, (raccoon, fox, rabbit, mink and muskrat,) for the Charleston market. In the same time boxes of screw-augers, invented and made by Benj. Evans, (at the place now owned by John G. Davenport,) and, after Evans' removal to Ohio, made by Joseph Smith and John Edmunson, were frequently sent. Cotton began to be sent by the load, in round bales, about the year 1801. After the Quakers left Bush River, (say after 1806,) very little flour, butter, beeswax or skins found their way to Charleston. I often recur to that period-- when Newberry was covered with small farms, when each homestead furnished pretty much the means of food and raiment -- and fancy that the people were then happier than they ever have been since.

A recollection of an incident in the beginning of 1802, I may, perhaps, be pardoned in repeating. A very large poplar tree lay at the mouth of the first branch, north of Hugh O'Neall's mills. Bush River was in flood; the water had entirely submerged the mill-dam. Hugh O'Neall, William Barret and Levi Hilburn concluded that, with a common batteau and a rope, after the tree was cut loose, they could tow it down to the sawmill of the latter, opposite to O'Neall's mill. Accordingly, they succeeded in getting the tree loose, and in towing it, until they neared the dam. Then the force of the water carried them beyond their point; the tree, batteau and all passed into an eddy below the sawmill. To get it above the sawmill was the object. Hilburn was persuaded to get out on the log, and with a pole force it along; the other two were to manage the batteau and tow. Having accomplished the most difficult part of the ascent, and reached a point where the water was deep, but comparatively still, the boatmen were continually calling out, "Pole, Levi; pole, Levi!" He, straining every muscle, made a mislick with his pole, and fell into water more than ten feet deep. Rising, he essayed to mount the log, but, it rolling under his hands, he received another ducking. At last he succeeded in mounting astride. Then again he was called on to "pole," but he swore one of his biggest oaths, (and anybody who ever heard Levi Hilburn swear must know it could hardly be excelled,) "that he would pole no more." Just then Barret, looking around at him, dripping, and with his usually large lips much swelled, said to Hugh O'Neall, "Did you ever see anyone look so much like Tom Lindsey's Nero?" The. name thus given adhered to him ever after. The poplar tree thus obtained was sawed into planks, and out of them were made the coffins for the two sons, the wife of, and Capt. Daniel Parkins himself, who died in the great epidemic of 1802, as detailed in No. 11.

In February, 1803, was the greatest snow ever seen in this State, unless it may be that that of 1851 equalled it.

In 1804 Hugh O'Neall, alone, began the mercantile business, and continued it until1809. Until the close of 1806 it was manifest that he was doing an excellent business. But the two dread enemies of a mere merchant, universal credit and the use of intoxicating drink by the merchant and his customers, were sapping the foundation of prosperity, reason and happiness.

I may be permitted *here* to say, that then, for many years previous, and for the fourth of a century since, every merchant sold, with groceries and dry goods, intoxicating drink by the "small." Everyone drank more or less; the morning bitters. the dinner dram, and the evening night cap were universal. Rum, (Jamaica, West India and New England,) was then almost entirely sold and drunk in stores. Whiskey belonged to the distilleries.

The use of intoxicating drink grew upon Hugh O'Neall, until, like Nebuchadnezzer, the judgment of God was upon him, and he was deprived ofthat which distinguishes a man from a brute, *his reason.*. This sad result, however, was not the work. of an instant; his habit of drink had made him negligent of his business and over-confident in cotton speculation. When the embargo of 1808 came upon the country he had in store with the Messrs. Bulow more than two hundred bales of cotton. He was largely their debtor, and he had authorized them to sell as they saw fit. Frequent attacks of mania *a-potu* foreshadowed the event. His son, a stripling of sixteen in 1809, ventured to ask him to abandon the cup. He made the attempt, but too late. Madness had already laid its iron hand upon him. He was a maniac. His cotton was sold at an immense sacrifice, his debtors were, many of them, insolvent, his creditors pressed their debts into judgments, his property was sold, and his wife and children turned out to shift for

Often has the writer seen his honored father caged like a wild beast; often has he seen him when it was dangerous for anyone to approach him. For four years this was his unfortunate state.

Reader, stop and think! Has not the writer cause to hate the traffic in intoxicating drink? Ought he not to pursue it to its destruction? May not his case be yours? May not you suffer as he has done? Let me entreat you -- let the truth teach you -- let others' sorrows learn you wisdom.

In 1813, July, Hugh O'Neall was restored to his reason, and, like Nebuchadnezzer, he gave God all the glory! Not a shade was left upon his mind; his memory, wonderful as it was before his insanity, was just as perfect after his recovery. He became a Friend in reality, as he had been raised in profession. No humbler, better Christian ever stood before his Master.

He set himself most diligently about repairing the wreck of his fortune. He gathered up much that was apparently lost, and paid many of his creditors, *those* who most needed it. He made three trips to Ohio, Indiana, Kentucky and Tennessee. His descriptions of the countries which he visited, the people whom he saw, and especially his accounts of his visits to his relations, were most felicitous.

In 1815 he determined never to drink intoxicating drink, and to his death, in 1848, he faithfully maintained his resolution. In August, 1820, he became a member of his son's family, and there, *as afath er,* he remained until his Father called him home.

He never desired or sought office. He was a Commissioner of Public Buildings from '99 for many years; he was a Commissioner of Free Schools from 1822 until he declined to serve longer.

In the unfortunate political schism, called Nullification, he was against it, and openly maintained the principles of the Union party. Like the venerable mother of Senator Butler, he could have said, as she did when secession was the prevailing sentiment of South Carolina, "I have seen two wars, and I never want to see another."

Hugh O'Neall's family consisted of one son, John Belton, four daughters, Abigail, (now Mrs. Caldwell,) Rebecca, who died in 1854, Hannah, who died in 1815, and Sarah Ford O'Neall.

Hugh O'Neall was not only gifted with a most superhuman memory, but he also possessed an excellent judgment and a clear and easy elocution. He was one of the kindest and most benevolent of men, and yet his sense of justice and right was such, he never, (after his recovery,) suffered his feelings to lead him astray.

In person, he was remarkable for a strong, vigorous, compact frame. He was five feet ten inches high; his head was a fine one; his hair receded on each side, leaving a high, intellectual forehead fully developed; his hair was thin, soft and silky, and perfectly black in his manhood; in age it was sprinkled with gray, still, however, leaving the black predominant. His eyes were blue, his nose long and Roman, his mouth was full and well formed. He died Wednesday, 18th October, 1848, about 2 P.M., having lived two months and eight days beyond eighty. one. He left surviving him his wife Anne, who on Friday, the 5th October, 1850, at ten minutes after 10 A. M., followed him to the silent house, having lived two months, wanting seven days, beyond eighty-three. His son and two daughters still remain.

In the progress of the war of 1812 everything became exceedingly high. When I use the word "high," I would not have you suppose that I use it in the sense of "tall," but in the meaning of "dear," or "costly."

Flour was a scarce article, selling readily at ten dollars to twelve dollars per barrel. The ladies at that time made *cakes* thin, and rather a holiday affair. Such a thing as using a whole barrel of flour in pound-cakes would have been regarded then as an astounding act of extravagance. I remember well, in 1816, bearing an old lady, who was seated at a table soon to be graced by a bridal party, as she was treated to a bit of pound-cake, say to the lady of the house, "It is mighty good, but mighty costly, though."

Near forty years of peace and prosperity have seen what was then a straggling village become a town, along whose western limbs daily speeds the iron horse, fed upon wood and fire, and drinking *naught but cold water,* bearing by his superhuman strength the trade and travel of our backwoods, and outstripping the wind in his flight from point to point, and have made us forget the use and wholesome economy of our ancestral homes.

As illustrative of the past, I recall an incident which occasioned much merriment when it occurred

It will be remembered by those who know anything of the history of South Carolina, (though I confess there are few who can penetrate the dark veil of the lack of information which hangs over her history,) that General Joseph Alston was the Governor from December, 1812, to December, 1814, two dark years of the war.

In that time it frequently became necessary for orders to be borne to the militia. The *post*, now commonly called the mail, came then slowly dragging itself along on horseback. The great Western mail passed then once a week on horseback, under the riding of the late Mr. Waddell, of Greenville. The orders of the Commander-in-Chief could not be allowed thus tardily to travel He sometimes sent *an aid.* The person who acted on the occasion to which I am about to allude was a Dominie Sampson sort of man, though not at all of his size, nor of his ungainly deportment. He was, or rather had been, however, a schoolmaster, private tutor -- tutor *pro tempore* in college, and thus *his fitness* for private secretary and aid, or anything e lse in the shape of man of business for the Governor is shown.

General Samuel Mays, of Edgefield, then commanded the first brigade. For some cause (perhaps in the absence ofthe Major General, Butler,) he was waited upon by the gentleman whom I have described. The General was not at home when he called. His kind, excellent lady invited him to stay until he returned. In the mean time, (as the family dinner had passed,) a dinner was provided for the traveler. Flour had of course to be put in requisition for the Governor's aid, but, guided by the precious character of the article, the cook made the biscuits small, very small. Dinner was announced. The hungry guest was paying his respects to the real good Carolina dinner, over which the General's lady, with hospitable intent, presided. A little black boy waited; his was the duty to hand the biscuits. The famished aid devoured a biscuit at a mouthful, and called to the waiter: "Biscuit, boy!" The little negro could not bear such wholesale destruction of his mistress' good things, and addressed himself at once to her. "La, Misses," said he, "he has had six already: shall I give him another?"

THE LAST QUAKER MEETING.

The cold, gray sunshine of an October Sabbath morning, preceding the bright gorgeousness of the Indian summer, seemed appropriate to the invitation I received to accompany a dear lady friend to the last meeting which has been held by her sect at the Quaker church at Bush river, Newberry district, South Carolina. Two Friends, an aged lady and gentleman, had come from a distant land on a visit to the few who remained of their persuasion, and to look upon the graves of all who had so peacefully departed to the blessed home of rest. The venerable Hugh O'Neall, whose striking biography appeared last week in the local district newspaper, and his aged companion and youngest living daughter, were all who remained of that people who once, with the olive branch of peace and industry in their hands, made the rich lands of that section of the district smile with their examples of thrift and economy. As we rode gently along, I had ample leisure to reflect upon the many social mutations which have already swept over our land in her brief period of national. infancy. We overtook the good old Father O'Neall a short distance from the church, mounted on his drab-colored pony, and looking like Old Mortality striving to defy time-- that silently moving power which carries every thing into nothing. Whosoever looked on that good man, in the over-ripe maturity of a virtuous old age, loved him. With a cheerful word and a heart-illuminating smile for all, he was the practical example of purity and elevated virtue. Rest there, old fathers, in thy quiet graves. The roaring winds of this wintry storm disturb not thy slumbers to-night, for thou wast with peace, beloved by God and by man.

The plain Quaker carriage of the visiting friends stood before the churchyard, and they were walking in silent meditation amongst the carefully heaped-up mounds which pious devotion had preserved from common disorder and neglect. It was a picture which, since then, has dwelt with me, and one which I have often thought I would pen-paint, that others might receive the satisfaction which the touching spectacle afforded. I was a boy then-- ambitious ofthe future-- with the world spread out before me; and since, its trials, its disappointments, its vexing cares have beset my path. But that day, and its impressions, have dwelt in the chambers of memory-- pure as a strain of music floating over distant waters. The gray old church, with its plain exterior, the singular garb of the pious Friends, the neatness of all the mounds -- even those of nearly a hundred years -- the bright colors of the dying leaves, already tinted by the autumnal frosts, were grouped into the picture, whilst the now mellow sunshine, reflected from the blue sky, draped it with beauty beyond the achievement of the pencil of art. The glory of that day's sunshine was God's smile upon the remnant ofhis children of peace. Silently, and one by one, as messengers from another land, they entered the church, and I felt at first that my presence might be an intrusion, where all was love and holiness; but the youngest, my lady friend, quietly bade me enter. We sat long and in meditation. Patience and meekness and long-serving and humility were thus silently taught to the hundreds who lay around in the peaceful slumbers of death; and the reflections which arose from the shrines of the past told the history of bygone years more eloquently than living words could have done. A cardinal red bird came and twittered among the delicate boughs of a red-fruited tree which grew over a grave, and its scarlet garb and shrill electric notes frequently, and for a long time repeated, were strangely contrasted with the quiet scene around.

Note after note he poured forth from his full-throated beak, whilst his swelling crest, and gay out-stretched wing, and voice of song, plainly told that he too was praising God in the bird recitative of nature's music. The aged mother arose, and the prose-voice of song in the mellow cadences, uttered in unison with the feelings of her heart, spoke of those who had passed away to light and peaceful glory in heaven. Whilst her words of love were poured out to the living and the dead, I fancied that one from another world, and from a long past age, was speaking. The old gentleman, with a clear, singing, mellow tone, then asked the empty seats and silent walls where those were who once peopled them. He bewailed the desolation in Israel, whose glory had departed, and whose land was peopled with strangers to the faith of their fathers. To me his words were as the lamentations of a second Jeremiah, saying: *"Our inheritance is turned to strangers, our house to aliens."* Again a brief silence: then the stillness is broken, and the voice of Hugh O'Neall, tremulous with emotion, tells the sad story of that faith by which he lived, and which, since then, made his dying bed a pathway of blessed ease, going home to God. The red mounds told the fates of many-over the blue mountains, beyond the broad Ohio -- others had fixed their homes in the wilderness, nearer *to* the setting sun. He and his alone remained -- here he had lived, and here he would lay down to rest in the grave. He said, still the seed of the faith was alive, for *"Thou,OLord, remainestforever,· thy thronefrom generation to generation. Turn thou us unto thee, O Lord, and we shall be turned; renew our days as of old."* I believe these words of eloquent lamentation from my aged friend were the last uttered in that silent house of God. Angels led out that little band of the true and faithful, and the sacred doors were closed forever. As we departed, the red-bird glanced through the tree-tops and chirped us a good-bye.

Death has since claimed all of those beloved Quakers save one, and may she long be spared to reflect the virtues of her heart in that social sphere in which she is a blessed and blessing visitant.

John Belton O'Neall
1793-1863

Judge O'Neall was born on 10 April 1793 in Bush River, Newberry District, South Carolina the son of honorable, debt-paying Quaker parents, Hugh and Anne Frost O'Neall.

He was a birth-right Quaker, but became a Baptist. On 24 June 1809, John Belton O'Neall was received in membership in minutes from New Garden Quarterly Meeting, because Bush River Monthly Meeting had been "laid down" following massive migration of Friends to the slavery-free North. He took in the children of the deceased Henry Miles O'Neall and his wife as his wards.

Educated in local schools and South Carolina College (second honor graduate, 1812), he studied law under John Caldwell and Anderson Crenshaw and was admitted to the practice of law and equity in May, 1814.

On 25 June 1818, he married Helen Pope, daughter of Captain Sampson Pope and his wife, Sarah Strother Pope of Edgefield. He died on 27 December 1863 in his house at Springfield, near Newberry, at age 70. Helen Pope was born on 19 November 1797 in Edgefield County, South Carolina. She died on 10 Novembr 1871 in Springfield (family house), Newberry District, South Carolina at age 73. The family house, Springfield, burned a few years after her death.

Judge John Belton O'Neall and his wife, Helen, are buried in the Rosemont Cemetery on Luther Street.

Judge O'Neall's remarkable personality made him a leader among men in various fields of endeavor -- Chief Justice of South Carolina, a judge for 35 years, a trustee of South Carolina College for 40 years, first President of the G&C Railroad Company, Speaker of the House of Representatives and Major General in the Militia. He excelled as a lecturer and as author of many articles and books, including The Annals of Newberry and The Bench and Bar of South Carolina. A staunch Unionist, he succeeded against keen competition, despite the unpopularity of his political views.

During a long and useful life, he contributed mightily to knowledge, jurisprudence, education, temperance, religion and agriculture.

John Belton O'Neall Plantation & Mill

Begin your tour on the Square in Historic Downtown Newberry! In 1826, Robert Mills described the village of Newberry as lying in the declivity of a hill (downward slope) about three miles from Bush River.

Head Weston Main Street. At the top of the hill to the left is the site of Newberry Cotton Mills. Begun in 1883, it was the first mill in America to be powered entirely by electricity. It was torn down in 1990. At the corner of Poplar Street on the left is Mayer Memorial Lutheran Church. On the left, following the north fork of Scott's Creek is Willowbrook Park. Turn left on Crosson Street. Behind Newberry Middle School is West End Cemetery. Although the mills were designed to be distinct communities, West End is the only one of the Newberry mill villages to have its own cemetery.

Tum right on O'Neall Street. When you cross Hwy 121, the road becomes Belfast Road. Somewhere off to the left, between Hwy 121 and Bush River is the site of Springfield, the home of John Belton O'Neall. In the early years of the nineteenth century, the farm had also been known as Kelly's Old Store and was the site of a store and mill in the eighteenth century.

John Belton O'Neall had about 50 slaves in 1850. They ranged in age from babies to 50 years old. He provided for his slaves and took care of them.

The Friends Quaker Bush River Church

The Quaker settlement was on Bush River and the Beaverdam. It extended from three to four miles on each side of the river. A line drawn from the Tea Table Rock, by the place once owned by William Miles, now the property of Mathias Barr, to Goggan's old field, now to Washington Floyd's, would be about the northwest limit. The settlement was prolonged down the river to the plantation, formerly the property of Col. Philomon Waters.

When the settlement commenced, cannot be determined with certainly. William Coate, before 1762, lived between Springfield and Bush River. Samuel Kelly, a native of King's County, Ireland came to Newberry from Camden and settled at Springfield in 1762. Benjamin Pearson and William Pearson lived on the plantation, once the property of John Frost, now that of Judge John Belton O'Neall as early as 1769. William O'Neall, father of Judge John Belton O'Neall, and family were early founders and members of the Bush River Quaker church.

The Friends had three places of meeting, one, the oldest and principal, at Bush River, where their house of worships still stands, neglected, but not desecrated. Within the graveyard, south of it, sleeps hundreds of the early settlers of Bush River.

Often over five hundred Friends, women and children, would gather to worship God in silence and to listen to the outpouring ofthe spirit, with which some of the Friends, male and female, might be visited. In this vicinage might be seen the brothers, Abijah, Hugh, William, John, Henry, and Thomas O'Neall.

In the beginning, Friends were slave owners in South Carolina. They, however, soon sat their faces against it, and in their peculiar language, they have uniformly borne their testimony against the institution of slavery, as irreligious. Such of their members as refused to emancipate their slaves, when emancipation was practicable in this State, they disowned. Samuel Kelly, who was the owner of a slave or slaves in 1762, when he came from Camden, refused to emancipate his on the grounds that he had bought and paid for them; they were therefore his property; and that they were a great deal better off as his property than they would be if free. He was therefore disowned. His brother's children manumitted theirs. Some followed them to Ohio; others have lived *here* free, it is true, but in indigence and misery, a thousand times worse off than the slaves of Samuel Kelly and their descendants. For the far-seeing old gentleman took good care in his last will, that the bulk of his slaves would be given to some member of or among his family.

Between 1797-1799, Abijah O'Neall and Samuel Kelly, Jr. bought the military land of Jacob Roberts Brown in Ohio. The great body of land was in Warren County, near Waynesville. Abijah O'Neall visited, located the land and in 1899, in the language of Samuel Kelly, Sr.

> Beyond the mountain and far away,
> With wolves and bears to play.

Abijah O'Neall commenced his toilsome removal to his western home. When about starting, he applied to the Friends for his regular certificate of membership. This they refused him on the ground that his removal was itself such a thing as did not meet their approbation. Little did they *then* dream that in less than ten years they would all be around him in the then far West!

Abijah O'Neall was about five feet eight inches high, stout, round-shouldered, light brown hair, eyes grey, nose Roman, mouth protruded slightly, his face had the appearance of great firmness. Such was his character. He was a surveyor and after he went to Ohio, spent much time in the woods in the pursuit of game. He was born in Newberry County, SC December 9, 1798 and died in 1874 in Crawfordsville, Indiana. He was a first cousin to Jolm Belton O'Neall. He was a member of the Indiana state house of representatives. He was raised a Quaker but later a Universalist. He sheltered escaping slaves as part of the "Underground Railroad" before the Civil War.

Between 1800 and 1804, a celebrated Quaker preacher, Zachary Dicks, passed through South Carolina. He was thought to have *also the power ofprophecy.* The massacres of San Domingo were fresh in memories. He warned Friends to come out from slavery. He told them if they did not, their fate would be that of the slaughtered Islanders. This produced in a short time a panic, and removals to Ohio commenced, and by 1807 the Quaker settlement had changed its population. Jolm Kelly, Sr., Hugh O'Neall, Jolm O'Neall, and Henry O'Neall were among those who stayed in South Carolina.

THE HOLLOWAY& PATTERSON FAMILIES

"4426 Cokesbury Road, Hodges, SC 29653
June 18, 2009

My grandmother was born in the 1800's. She was married to Rube or Ruben Patterson in the year 1900 (census shows 1899). Her name was Willie Ann Holloway and her father's name was Sam Holloway. Ruben's parents: Willis and Nancy McKensey. Siblings: Wince, Nelson, Jane, George, Ben; it is believed to have many more (approx. 18).

Her mother was Martha, from Queensland, Australia. She had sisters Emma, Mary, Martha, Caroline, Ada, Susa, and Adeline; she had more but I just don't have all of their names. Her brothers: John, Henry, Matthew, Morris, West, Sam, George, Edward, and Solan. It was a big family of them.

Their Daddy was a slave and named Sam after his father. His mother was a slave, also.

Dorothy A. Chappelle"

••

The following family members have been compiled from birth and census records. It must be remembered that Edgefield, Newberry, and Spartanburg Counties were once part of the Abbeville District. Before 1860, Negroes were listed (if listed) as Indians. It was 1870 and later before they were classified as Blacks or Mulattoes.

1870 Saluda, Edgefield, South Carolina census:

Willis Holloway	50 (b. abt 1820)
Caroline -wife	40
Sam Holloway	14
Wiley Holloway	22
Wade Holloway	24
Mary Holloway	18
Betty Holloway	8/12

1870 Ninety Six, Abbeville, South Carolina census:

Mary Holloway	30 (head of house)
Laura Holloway	14
Wash Holloway	12
P. Holloway	9
Jane Holloway	7
Fannie Holloway	6
Rody Holloway	1

Sam and Martha (Folks) Holloway Families

1870 Blocker, Edgefield Co., South Carolina census:
Luke Holloway, age 41; Rose Holloway, age 33; Roda Holloway, age 23, Paul Holloway, age 19; John Holloway, age 16; William Holloway, age 10; Samuel Holloway, age 5; Nancy Holloway, age 4; Robert Holloway, age 2; Arthur Holloway, age 1; Luke Holloway, Jr. age 1.
1870 Shaw, Edgefield Co., South Carolina census:
Silas Holloway, age 45; Sarah, age 35; wife; Emmanuel, age 19; Will, age 35; brother; Wiley, age 12; Frances, age 2; Martha, age 8/12.
1870 Dyson Mills, Edgefield Co., South Carolina census:
Willis Holloway, age 50; Caroline, age 40, wife; Sam Holloway, age 14; Wiley, age 22; Wade, age 24; Mary, age 18; Betty (Elizabeth), age 8/12.

1880 Blocker, Edgefield Co., South Carolina census:
Harry Holloway, age 82, mulatto, born SC; father b. SC, mother b. VA; Sharlotte (Charlotte) wife, age 72; Celia Brooks, age 17, daughter.
1880 Ryan, Edgefield Co., South Carolina census:
Adeline Holloway, born Jan. 1845; married Miles Parks (b. 1827); Nancy Mims, age 7, stepdaughter; Robert E. Parks, age 1, son.
Watson Parks, age 26; Adeline, age 25, wife; Thomas, age 4; John, age 3; Rosanna, age 2; Washington, age, 1/12; Lewis Holloway, age 48, brother-in-law; Virgis, age 17, brother-in-law; Mary, age 14, sister-in-law.
On same sheet: Samuel Mims, age 45; Susan, age 30, wife; Emeline Parks, age 13, stepdaughter, Elizabeth Parks, age 9, stepdaughter.
Peter Holloway, age 42; Mandy, age 35, wife.
Pink Holloway, age 22; Jennie, age 20, wife.

1900 Plum Branch, Edgefield Co., South Carolina census:
Adeline Parks, age 55,widow; Robert, age 23; Gary, age 18; Felix, age 15. Adeline died 12 November, 1917, age 75.
1900 Bordeaux, Abbeville Co., South Carolina census:
Dave Holloway, age 40; Mary, age 26, wife; Rosa, age 8, Luce E., age 8; Jacob, age 3; Mamie, age 11/12; Willis, age 15; Lewis, age 11; Forster, age 8.

1920 Brooks, Greenwood Co., South Carolina census:
Elijah (Elizah) Holloway, age 45;Willie, age 19, wife; Henry, age 18; Alma, age 14.
Henry Holloway, age 50; Winnie, age 32, wife; Ozie, age 7; Quincy, age 6; James, age 5; Wesley, age 4; Welton, age 2 9/12; Willie, age 1.
John H. Holloway, age 36; Alma, age 22, wife; Henry, age 4 8/12; Sloan, age 2 5/12; Masie, age 5/12.
1920 Colliers, Edgefield Co., South Carolina census:
Lewis Holloway, age 41; Mary, age 39, wife; Martha, age 16.
Wesley (West) buried. Graniteville, Aiken Co., SC; Matthew b. Enoree Zion Baptist; Morris b. Enoree Zion Baptist; Sam b. Mount Moriah, Greenwood; Emma b. Mount Moriah, Greenwood; Willie Ann b. Beulah Baptist Church; Ada, Greensboro, NC; Susa b. Enoree Zion Baptist; Mary b. Mt. Tabor.

1870 Hellers, Newberry County, South Carolina census:

Thos W. Holloway	40	(slave passing as white?)
Augusta Holloway	29	
J. B. Oniell Holloway	14	
Willie R. Holloway	11	
Chicora Holloway	8	
Mary E. Holloway	5	
Nanna Ramage	15	
John S. Hobb	12	
Jeff Davis Hobb	10	
Peter Niles	17	
A. Yarborough	50	

AJ Holloway	20	farm laborer
Lizzy	19	domestic

1880 Newberry, Newberry County, South Carolina census:

Caroline Boozer	50	
Riley Holloway	15	(death certificate shows his mother as Caroline Harris and his father as Washington Holloway)
Wash. Holloway	13	

1880 Newberry, Newberry County, South Carolina census:

Sam Veal	27	
Mary Veal	23	(Mary Holloway of slave interviews)
Addie Veal	8	
Jim Veal	4	
Sallie Veal	2	

Birth certificate shows Joseph Veal, born April 11, 1885, to Mary Moore and Samuel Veal.

1900 Greenwood, Greenwood County, South Carolina census:

Sam Holloway	46	
Emma Holloway	15	
Sam Holloway	6	
Blik Johnson	30	(boarder)

On the same sheet:

Laura Holloway	40	widow
Otavia Holloway	38	single

1900 Kinards, Greenwood County, South Carolina census:

Mat Holloway	36	(Matthew)
Lila Holloway	28	

Ed Holloway	15
Mannie J. Holloway	12
Wesley Holloway	8
Emma Holloway	6

Death certificate shows parents as Henry Holloway and Pathnie, both born Ninety-Six District.

1920 Greenwood, Greenwood County, South Carolina census:

Henry Holloway	50
Minnie Holloway	32
Ozie Holloway	7
Qunecy Holloway	6
James Holloway	5
Wesley Holloway	4
Melton Holloway	2
Infant	0/12

1920 Greenwood, Greenwood County, South Carolina census:

Sam Holloway	69
Alie Holloway	68
Boame Holloway	13
Laura Holloway	4
Infant	0/ 12

1920 Brooks, Greenwood County, South Carolina census:

Elizah Holloway	45
Willie Holloway	19
Henry Holloway	18
Alma Holloway	14

1920 Hodges, Greenwood County, South Carolina census:

Tom Holloway	24
Bessie Holloway	38
Lula Holloway	11
Lillie M. Holloway	9
Johnie Holloway	1
Infant	8/12

1920 Greenwood, Greenwood County, South Carolina census:

Nathan Harris	50	
Annie Harris	49	
George Holloway	26	stepson
Lula M. Holloway	13	stepdaughter

1920 Kinards, Greenwood County, South Carolina census:

Morris Holloway	46

Sallie B. Holloway	26	
Ben Holloway	24	
Mattie M. Holloway	2	
Infant	1/12	
Lizzie Pressley	13	

1920 Greenwood, Greenwood County, South Carolina census:

Mehaly Holloway	65	widow
May Holloway	33	single
Annie M. M. Holloway	5	granddaughter
Sam Lee Holloway	3	grandson

1930 Cokesbury, Greenwood County, South Carolina census

Sam Holloway	62
Mimie Holloway	48
Joseph Holloway	17
James Holloway	6

1930 Greenwood, Greenwood County, South Carolina census:

Henry Holloway	69
Minnie Holloway	47
0. C. Holloway	16
Quincy Holloway	15
James Holloway	12
Wesley Holloway	10
Fannie Holloway	9
Charlie Holloway	8
Will Holloway	6
Dusk Holloway	4
Infant	1/12
Susie M. Holloway	3
Infant	2/12

1880 Hibler, Edgefield County, South Carolina census:

George Patterson	42
Rachael Patterson	40
Seymour Patterson	12

On same page:

Polk Hollaway	34
Susan Hollaway- wife	27
Harriet Hollaway	7
Henry Hollaway	2
Infant daughter	1/12

1900 Fellowship, Greenwood County, South Carolina census:

Rube Patterson	59
Willy Ann Patterson	25 (given as wife)
George Patterson	6/12
Will Moore	12 (given as boarder?)

Delayed birth certificate for Will Patterson (Will Moore?), born Feb. 25, 1888, parents given as Reuben Patterson and Willie Ann Holloway.

1910 Greenwood, Greenwood County, South Carolina census:

Rube Patterson	32
Willie Ann Patterson	30
Carrie Patterson	9
Mamie Patterson	7
Willie Patterson	5
Mose Patterson	4
Ruben Patterson	2
Ebbie Patterson	3/12

On same page:

Reuben Holloway	24
Emma Holloway - wife	23
Evander - son	4
Annie Joe - daughter	2

1920 Cokesbury, Greenwood County, South Carolina census:

Reuben Patterson	44
W. A. (Willie Ann) Patterson	40
Louise Patterson	18 daughter
M. F. Patterson	14 daughter Mamie L.
W. R. Patterson	14 Willie Robert
Rock Patterson	13 David
Reuben Patterson	10
T. L. Patterson	9 son Robert
L. S. Patterson	7 son Lewis
W. M. Patterson	5 daughter Willie Mae
Edward Patterson	2 son
Infant	4/12
Elva Patterson	6/12 daughter Glenen
Malik Jones	13 niece Mae Bell

1930 Greenwood, Greenwood County, SC census:

Reuben Patterson	65
Willie A. Patterson	50
David Patterson	24
Reuben Patterson	21

T. L. Patterson	20	son
L. F. Patterson	18	son
William F. Patterson	15	
Eddie Patterson	12	
Evelyn Patterson	10	
Cora I. Patterson	7	
Kathleen Patterson	10	
May B. Jones	24	niece

Willie Ann Patterson is buried at Beulah Baptist Cemetery, Greenwood, SC.

Her sister, Mary Holloway Williams is buried at Mt. Tabor Church Cemetery, Greenwood, SC.

Adeline Holloway, daughter of Sam & Martha Holloway married Miles Parks, age 53 on 1880 Edgefield Co., SC census. Stepdaughter, Nancy Mims; son Robert E. Parks. 1900 Edgefield Census: Adeline, widow, age 55; children: Robert, 23; Gary 18; and Felix 15. Adeline died 12 Nov. 1917, age 75. On same census: Watson Parks, age 26; wife Adeline, age 25; Thomas age 4; John age 3; Roseanne age 2; Washington age 1/12; Lewis, age 48, Girgis, age 17, brother in law?; Mary, age 14, sister-in-law.

1880 Edgefield Co., SC census
Peter (Niles) Holloway, age 42, wife Mandy, age 35
Nancy Holloway, age 50; sister Jane, age 49

1870 Blocker, Edgefield Co., SC census:

Luke Holloway	41	
Rose A. Holloway	33	wife
Roda Holloway	23	dau
Paul Holloway	19	son
John Holloway	16	son
William Holloway	10	son
Samuel Holloway	5	son
Frances Holloway	2	dau
Robert Holloway	2	son
Arthur Holloway	1	son
Luke Holloway	1	son

Harry Holloway	73	(father born SC; mother VA)
Charlotte Holloway	65	

History of Williamsburg, From 1705- 1923, By William Willis Boddie
Columbia, SC; The State Company, 1923
Transcribed by Dena W. for Williamsburg County, South Carolina Genealogy Trails

CHAPTER III.
ORIGINAL SETTLERS.

From 1735 to 1737, a great many settlers came to the new township on Black River and practically every acre of land had been taken up by these settlers within a year after the township had been surveyed. Every man settling here was granted a half acre town lot and fifty acres of land in the township for himself, his wife, and each one of his children.

These are the names of the heads of families who had settled in Williamsburg Township up to 1737: Robert Allison, John Anderson, James Armstrong, David Arnett, James Adams, John Athol, John Ballentine, John Barnes, George Barr, Joseph Barry, John Basnett, Benjamin Bates, Matthew Bernard, Joseph Bignion, James Blakely, John Blakely, John Bliss, John Borland, Jonathan Bostwick, James Bradley, Thomas Brown, George Burrows, William Camp, William Campbell, William Cochran, John Connor, William Copeland, William Cooper, James Crawford, Thomas Dale, John Dick, Nathaniel Drew, Thomas Dial, Robert Ervin, Francis Finley, Robert Finley, James Fisher, John Fleming, John Frierson, William Frierson, Aaron Frierson, David Fulton, James Gamble, Roger Gibson, Gabriel Girrand, John Gotea, Roger Gordon, Francis Goddard, Hugh Graham, Hugh Green, George Green, Richard Hall, Thomas Dial, Archibald Hamilton, William Hamilton, Christopher Harvey, William Harvey, John Herron, George Hunter, Jeter Hume, John James, William James, John Jamison, William Johnson, Joseph Johnson, David Johnson, Abraham Jordan, Samuel Kennedy, John Knox, Crafton Blerwin, Richard Hake, John Lane, James Law, Patrick Lindsay, William Lowry, Richard Malone, John Matthews, Samuel Montgomery, Daniel Mooney, John Moore, William Morgan, Joseph Moody, John McCullough, Nathaniel McCullough, Daniel Murray, David McCants, John McCants, James McCauley, James McCutchen, James McClelland, Alexander McClinchy, William McCormick, William McKnight, John McElveen, Thomas McCrea, Alexander McCrea, William McDole, Hugh. McGill, David McEwen, James McEwen, Andrew McClelland, James McGee, Edward McMahan, Matthew Nelson, John Nicholson, William Orr, James Pollard, John Porter, John Pressley, William Pressley, Edward Plowden, John Robinson, Joseph Rhodus, Andrew Rutledge, John Scott, James Scott, William Scott, James Smith, Charles Starne, James Stuart, John Stubbs, John Sykes, William Syms, James Taylor, William Turbeville, William Troublefield, Matthew Vannalle, John Whitfield, William Williamson, Hemy Williams, Anthony Williams, David Wilson, John Wilson, William Wilson,

David Witherspoon, Gavin Witherspoon, James Witherspoon, John Witherspoon, Robert Witherspoon, Robert Wilson, and Robert Young.

These original settlers in Williamsburg Township came from England, Ireland, Scotland, Germany, Holland, and from the New England States, Pennsylvania and Virginia. They were all about the same class of men. They were people who had been non-conformists as to State-Church religion, and nearly all of their families had lost their property in the religious conflicts of the seventeenth century. The greater number of them had lived in Ireland for many years before coming to America. They had migrated :from England and from Scotland to Ireland on account of fair promises on the part of the English King. These failing them, they sought refuge in America.

The Blakelys, Bradleys, Browns, Finleys, Gambles, Halls, Humes, Johnsons, Matthews, Hurrays, Nelsons, Plowdens, Rutledges, Taylors, and Wilsons were of English blood. The Barrs, Dials, and others were of German descent. The Bignions, Janneretts, Vanalles, and Orrs were of Swiss origin. The Barrys, Kennedys, Lindsays, Lowrys, Malones, and Morgans were Irish. The Arnetts, Campbells, Crawfords, Ervins, Friersons, Fultons, Flemings, Grahams, Hamiltons, Montgomerys, McColloughs, McCreas, McGills, Pressleys, Scotts, and Witherspoons were Scotch-Irish. The Williams and the James families were Welsh.

Page 84:

John Fleming's will was proved before James McCants, Esq., May 11, 1768. He mentions his wife, Elizabeth; his brother's daughter, Elizabeth; his cousin, Samuel Shannon; his sister, Agnes Cooper alias Fleming, and her two sons, James and Thomas Cooper; his sister's son, George, and daughter, Elizabeth Cooper; his wife's daughter, Jannet, and her daughter, Elizabeth Blakeley; his brother, James Fleming and his son, Peter Blakeley Fleming.

Page 185:

There were several Presbyterian churches in the territory surrounding Williamsburg that had considerable influence on this district during the period between 1780 and 1830. These congregations had been organized by people migrating from Williamsburg; and not withstanding the condition of what they called roads, kept up communication with their friends and relatives in this district. Of these churches, Salem Black River in the Sumter District and Hopewell and Aimwell on the Pee Dee had been founded between 1760 and 1770 and were large and aggressive churches. They were about 40 miles from the King's Tree. This great distance meant much at that time, although the congregations of these churches usually came to Williamsburg to attend spring and fall communion services and camp meetings.

In September, 1801 John Witherspoon, John Witherspoon, Jr., Archibald Knox, William Mcintosh, Thomas Rodus, Daniel Epps, John McFadden, Thomas McFadden and Samuel

Fleming met at the home of Mrs. Mary Conyers, who lived about half way between the Williamsburg Church and the Salem Black River Church, and organized a Presybyterian Congregation for their community. John Witherspoon, John Witherspoon, Jr., and Archibald were named its first elders. On November 10, 1802, the building was completed and called Midway because it was halfway between the two well-known churches just named. The Reverend C. G. McWhorter gave one-fourth of his time to this new church.

In 1809, Midway had twelve members. That year the Reverend John Cousar preached two Sundays every month at Midway and the other two at the Brewington Church. Midway Church is located on the northeastern branch of Black River in what is now Clarendon County.

Page 332-334:

Slavery was the source of great power in the South. The North realized this. It is but a short way from the realization of an economic interest to the actualization of a strong religious sentiment to sustain and support it. It did not, therefore, take a long time for professional Puritan religious reformers to begin to preach and to pray about the evils of African slavery in the South. Some of them were sincere and honest, perhaps most of them, but like all reformers, they lost themselves in their own delusions. The power which slaves gave the South in Congress was the real reason for so many pathetic "Songs of Labor" which were written in the section where no darkies sang around the "Great House" door. The rabid Abolitionists at the North usually proclaimed only the inhumanity of African slavery in the South. The most charitable thing that may be said about them is that they did not know.

There is no defense of African slavery in the South. African slavery in the South did not grow out of missionary ideas for the promotion of the Christian spirit in the world, nor was it designed for the promotion of altruistic sentiment, but it did take a race of men that for thousands of years had roamed the pampas and plains and jungles of Africa like wild beasts, and, within a single century, bring forth multitudes of substantial Christian men and women.

In 1808, there was a shipload of Guinea negroes sold in Williamsburg, South Carolina. They spoke no language save that of grunts and nods. They know not their right hand from their left. One hundred years later, in 1908, the descendants of these same Guinea negroes were prosperous citizens. Some of them owned considerable plantations and produced large crops of com, cotton, and tobacco; some of them were members ofthe bar in New York City; some were practicing medicine; some were architects and builders. There is no other case of such remarkable development of a race recorded in history or told in tradition.

Slave owners in Williamsburg, South Carolina, were neither more or less saintly, human nor inhuman, than other men of the world of their day and generation. When these savage Africans were brought to this district, they could not be disciplined or controlled and civilized and made servicable only by the use of smooth tongues and gloved hands. It was necessary that the planter transform a wild man into a profitable workman within a short period oftime that the slave might be profitable. This was done well done with as little physical force as was expedient....Slaves were required to render instant and unquestioning obedience and this proved their salvation. Out of this slave training, came some of the most noble characters, the most loyal subjects, and the most beautiful service ever seen in the world. It required clear minds, strong arms, and end-less patience to make Guinea negroes into servicable citizens. South Carolina did it.

Bill was the son of a negro captured in the jungles on the Congo, and sold as a slave on the block in Charleston. In the graveyard about the beautiful old Black Mingo Baptist Church, one now finds a marble slab on which is graven: "Sacred to the memory of Bill, a strictly honest and faithful servant of Cleland Belin. Bill was often intrusted with the care of Produce and Merchandise to the value of many thousand dollars, without loss or damage. He died in October, 1854, in the 35th year of his age, an approved member of the Black Mingo Baptist Church. Well done, thou good and faithful servant. Enter thou into Joy of thy Lord."

It is told of a large slave owner, one in Williamsburg District, that sometimes he moved up and down his line of slaves, while they were working in the fields, beating and promiscuously with his cane, and sometimes knocking one senseless. It is told of another slave owner in the district that once he hanged a negro man up by the tumbs and used the claws of an enraged tom cat to lacerate the bare back of the suspended slave. These two stories are probably true. That same master of whom the story is told of suspending the negro by the thumbs was seen one morning digging a ditch in a swamp while several of his slaves stood near on dry ground. A man passing asked the master why he did not make the negroes do the work. The master replied, "It might make the negroes sick."

The McCants (also spelled McKantz) were early settlers of Williamsburg District, South Carolina as early as 1735-1737. James B. McCants was a prominent lawyer in the Fairfield District and was worth $35,000.00. The census does not show how many slaves he owned.

The two leading plantations where these slaves were located were: Cedar Hill Plantation, Abbeville, South Carolina, owned by John Calhoun and later by Thomas Pettigrew; Snee Plantation of Charleston, South Carolina, 1859 owned by William B. McCants.

In George Fleming's slave narrative, he said he married (1) Sallie Patterson, (2) Lonie Golding, and (3) Elizabeth McKantz (McCants). Elizabeth was from Abbeville, South Carolina, father a white man. She is probably the sister of Nancy. In Wm. B. McCant's will, he lists a slave Elizabeth for $250.00, two slaves named Nancy valued at $450 and $250 George valued at $600.00 George Fleming was the property of Sam Fleming. George's father was named Bill and stole from Virginia; his mother's name was Hannah and they were married on the plantation in Laurens County, South Carolina. Sallie Patterson was the daughter of George Patterson.

208 " " " Table to Bell Bentham
Two Tea Spoons to Mary Louisa McKinley
A lot Clothing to Caroline Plumeau

Box 34 No 18

A true and perfect Inventory of the Goods and Chattels of the Estate of Joseph B Whaley deceased as shewn by the Administrator W. B. Whaley. Depd 1 June 1846

Two Slaves Jubiter and Sandy valued at $450
House furniture valued at 500
Carriage and Sulky.......... 150
Four horses .... $ 300
Two Mules 100
Three Carts one waggon two Ploughs..... 75
Thirty one head of Black Cattle 124
Twenty two Sheep................ 27.50
$1826.50

W James Whaley Thomas A Baynard John L. Wescoat

Box 33 No 19

Inventory of Estate of R J La Roche of Wadmalaw Island appraised on the 13th and 14th May 1846 Depd 4 June 1846

Negroes on Wadmalaw Plantation
Manuel $300 Coty $200 Jack $400 ------ $ 900
Little Bristol $600 Cornelia $450 1050
L Sue $300 Old Jack $1 301
Jack $300 old Phillis $100 400
Bristol $200 ---- ---- ---- 200
Cuba $350 Billy $ 400 750
Anthony $600 Lizzy $ 450 Ned $50 1100
Jack $700 Chanty $450 Judy $300 Augustus
$250 Sally $200 Ephraim $150 Linda $50. } 2100
Jimmy $400 Lovey $475 Caty $450 Cupid $500
Joanna $300 Eve $250 Flora $150 Mingo $100. } 2625
Sue $200 Abram $100 Mary $250 Betsy 300
$400 Pompy $375 Carolina $300 Anne $200 } 1525
Tom $500 Beck $450 Edward $500 Margaret
$300 Nan $150 Eliza $100 Lucy $50 ---- } 2050
Lot $200 Little Lot $600 Venus $200 Fanny
$450 Bella $450 Emma $450 Lydia $150 Dell $[illegible] } 2600
Dick $450 Ido $350 Betty $450 Ezekiel $550 1800
L Manuel $600 Phillis $400 Annette $300
.... Joshua $75 } 1375
Frank $650 650
Joe $ 400 Charlotte $75 Isaac --- 475
Flora $450 Abram $400 George $400 Frederick
$250 Dick $200 .... } 1700
Fortune $600 Eubia $450 Jimmy $100 Maria $50
Joshua $600 ---- ---- } 1800
Frank $400 Cornelia $450 Susan $350 --

Elizabeth $250 Adeline $150 1600
Toney $600 Joan $450 1050
Lydia $500 Daniel $600 Clays $400 Mary $300 1800
Nanny $150 David $600 Robert $600 1350
Peter $450 Sary $800 Andrew $600 }
Cuffee $600 Nancy $250 Peter $175 } 2375
Sylvia $250 Amey $450 Caesar $450
Elsey $200 Isaac $300 Chloe $200 1850
Smoke $600 Bella $500 Clarissa $250 Titus $100 1450
Old Anthony $10 Catrina $10 Titus $600 Adam 600 1220
Anthony $600 Nancy $450 Mingo $600 1650
Yotter $700 Nanny $450 Moses $600 Charles
$600 Joe $550 Syllie $300 Elsey $450 Caroline
$100 Edward $50 Jimmy $50 3850
June $600 600
Jeffry $600 Minda $300 Dee $450 Jeffry $450 1800
Hannah $450 Maria $250 Yotter $175 875
August $800 Peg $200 George $600 Linda $450 }
Rose $450 } 2000
Warley $600 Doll $150 750
Peg Toney $600 John $600 Clara $450 }
Phillis $400 Sylla $200 .... } 2250
Ned $600 Nancy $50 Harriet $400 Tom $100 1150
Total amount of negroes $57321
1 Large Cotton Boat $250
1 8 Oared boat 200
1 4 D° D° 30
69 horned Cattle @5 345
70 " @6 420
75 Sheep .... @150 .... 112 50
1 Carriage & horses 275
5 Horses 150
House hold furniture 500
7 Carts 115
30 Gins 60
1 Horse power 50
Poultry 50
20 head hogs 50 2607 50
Total Amt property valued 53928 50
Wm McCants William Sams Benj Freeman
James and Benjamin Freeman's Bond
conditioned for $425
√ J J La Roche Sr Richd J LaRoche Jas La Roche

Sales at auction by order of the Administratrix
Box 28 of the Estate of the late Daniel Boinest May 15 1845
N° 17 Ligniez P 7 negroes to wit Pembroke Judy
and the five Children Stephen Frank
Matthias Daniel & Pembroke Junr $370 2590.

All *Slave Narratives* results for *Joe Crews*

Searching for...

Name: Joe Crews

Edit Search

or Start a new search

Narrow by Collection

All Categories

Stories, Memories & Histories

Family Histories, Journals & Biographies

Slave Narratives

ABOUT *SLAVE NARRATIVES*

Impressive collection of interviews of former slaves in the United States

Learn more about this database

Hot Keys

New search

Refine search

Preview current record

Highlight next record

Highlight previous record

Matches 1-10 of 20 Sorted By Relevance

State: South Carolina

Interviewee: Fleming, George

"Some men, like old Joe Crews, was reg'lar nigger traders. Dey bought niggers, stole 'em frum Virginia and places and drove 'em through de country like a bunch of hogs. Dey come in great gangs. In town dey have big nigger sellings, and all de marsters frum all over de countryside be dar to bid on 'em. Dey put 'em up on de block and holler 'bout dis and dat dey could do and how strong dey was. 'Six hundred --- Yip, Yip, make it six-fifty' I heard 'em call many times when I be dar wid Marse. Some of dem throw a thousand dollars quick as dey would ten at a purty gal. Some traders stop a drove of niggers at de plantation and swap or sell some. Dey didn't call dat putting 'em on de block like when day had de big selling.

View Full Context

State: South Carolina

Interviewee: Fleming, George

"Soon atter de war dar was a lot of trouble 'bout voting fer de governor. Some folks (like old Joe Crews) tried to put in de niggers' heads to vote fer de Republicans, but I know'd better. I voted fer Hampton like Marse did. Fact is, I voted twice fer him. (Joe Crews, and other scalawags like him,) Some scalawags had done made all de money dey could off selling niggers, so dey thought dey could make some more by making 'greements wid de Republicans. My daddy, Bill, was bullheaded. He done got dem ideas in his head and he said he gwine to vote fer de Republicans in spite of hell.

View Full Context

State: South Carolina

Interviewee: Fleming, George

Joe Crews told 'em dey couldn't bring 'em to de polls. He thought de Yankees would protect de niggers, but fact is, de Yankees done been paid off by de Democrats and left town. Us Democrats broke in de storehouse in Tin Pot Alley and got every one of dem guns. De niggers names was on de stock of de guns. We sho had a hot time when dem niggers come up dar trying to vote. Dat's when my daddy got kil't. He had already been shot in de leg befo' dat, and dey called him 'cripple Bill'. Dem was de purtiest guns I ever seed. Dey click three times when de trigger was pulled back. Old Jim Crews was kil't, too, at dat time. Wash Hill was de one dat got him. He was shot at Crew's Branch. Twan't long atter dat till things begin to settle down, fer de Democrats sho did lick up dem Republicans.

View Full Context

State: South Carolina

Interviewee: Patterson, George

He also stated that his father was a full-blooded Indian who was sold to his master by Joe Crews, the biggest slave trader in the country. His father was stolen somewhere in Mississippi, along with other Indians, and sold into slavery with the "niggers." He said his father told him he was stolen by Joe Crews when he was a young buck. At that time, his father went by the name of "Pink Crews," but after he was purchased by Mr. Joe Patterson, his name became "Pink Patterson." He stated that his mother was a white woman who came from Ireland and was working on the Patterson farm. She was not a slave, but was married to his father by his "Marster."

View Full Context

State: South Carolina

Interviewee: Patterson, George

"You remember reading about Joe Crews and Jim Young - what they did in this state? Well, they tried to lead all the niggers after the war was over. I was the one who got Jim Young away from the whites. I carried him to Greenville, but he got back somehow, and was killed. Joe Crews was killed, too. The

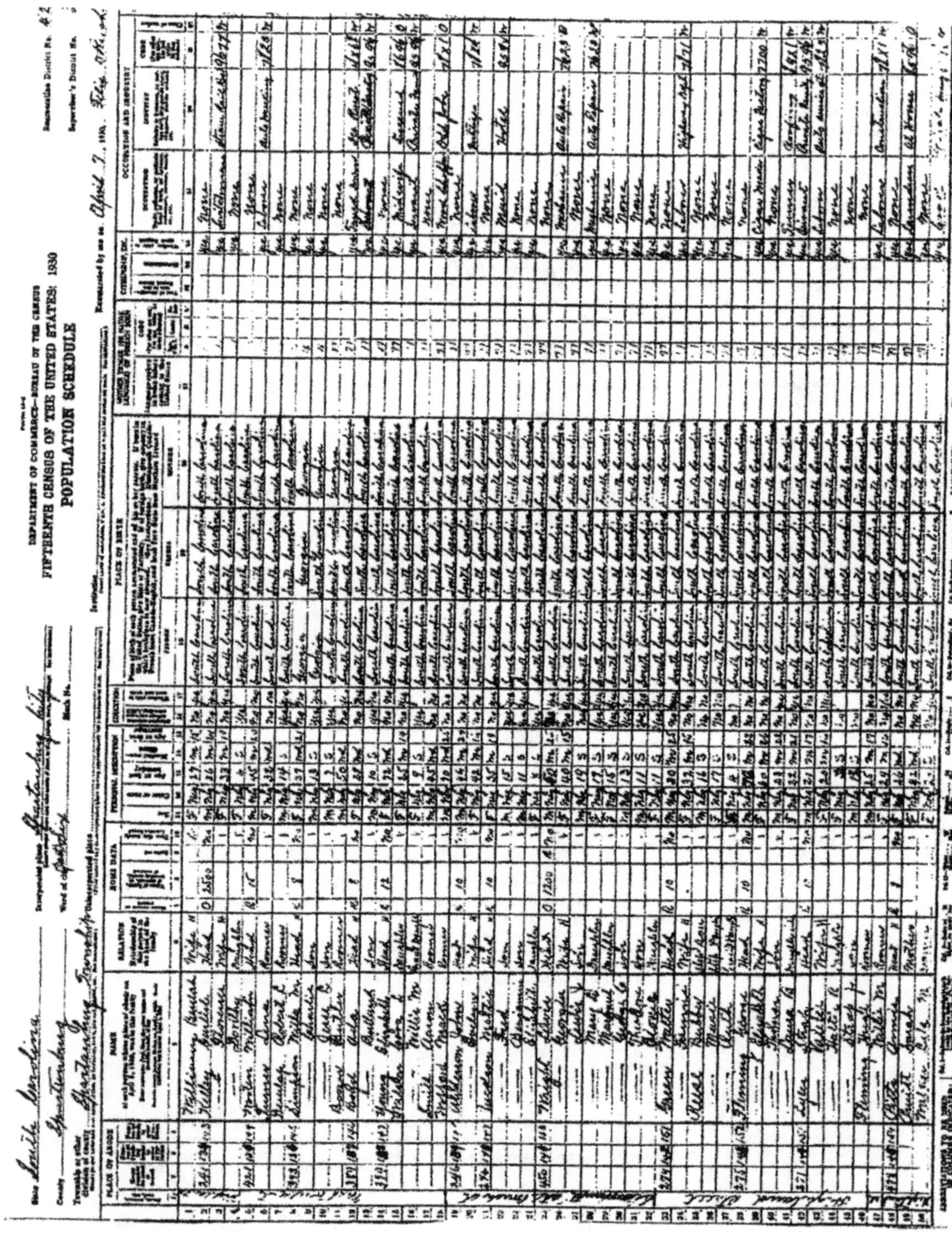
DEPARTMENT OF COMMERCE—BUREAU OF THE CENSUS
FIFTEENTH CENSUS OF THE UNITED STATES: 1930
POPULATION SCHEDULE

Ku Klux was after them hot, but I carried Jim Young away from them. You know, the Yankees was after getting all the gold and money in the South. After the war, some Yankee soldiers would come along and sell anybody, niggers or whites, a gun. They were trying to get on to where the white people kept their money. If they caught on, they would go there and steal it. You know, there wasn't any banks, so people had to keep their money and gold in somebody's safe on some big man's place. These men in selling guns was trying to find out where the money was hid."

View Full Context

State: South Carolina

Interviewee: Patterson, George

George also stated there were plenty of wild turkeys, ducks and wild geese on the Enoree River. The turkeys would ravage a garden or scratch up the planted seed on the plantation. He has often been sent out to frighten the wild turkeys away from the crops. He said plenty of meat could be secured by shooting the wild hogs that roamed the woods, that anybody was at liberty to kill a hog. Of course, some once tame hogs mingled with the droves of wild hogs but the tame hogs had the owner's name on them; so one had to be very careful that he did not shoot a marked hog. He said that when his father, an Indian, was stolen by Joe Crews, from the woods of Mississippi, he marched them with niggers he had also stolen, or traded for, into different sections of the country, selling them as slaves and speculating on them. He drove them just like cattle and would stop at various plantations and sell the Indians and niggers into slavery.

ancestry

[redacted]

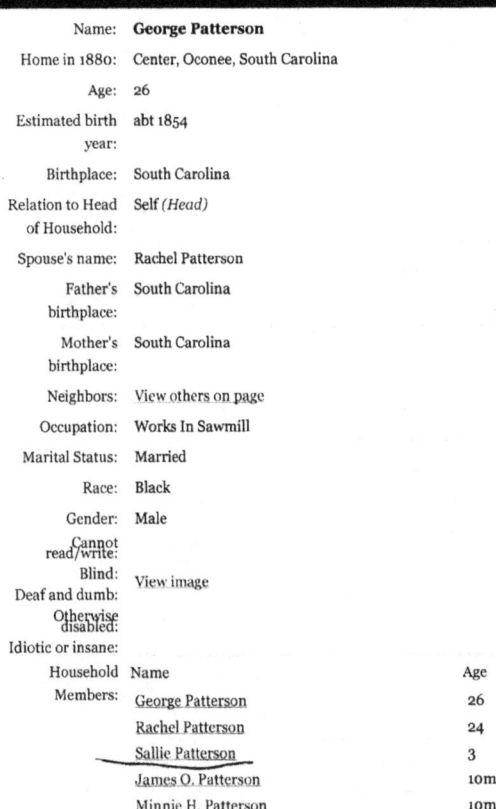

Name:	**George Patterson**
Home in 1880:	Center, Oconee, South Carolina
Age:	26
Estimated birth year:	abt 1854
Birthplace:	South Carolina
Relation to Head of Household:	Self *(Head)*
Spouse's name:	Rachel Patterson
Father's birthplace:	South Carolina
Mother's birthplace:	South Carolina
Neighbors:	View others on page
Occupation:	Works In Sawmill
Marital Status:	Married
Race:	Black
Gender:	Male
Cannot read/write:	
Blind:	View image
Deaf and dumb:	
Otherwise disabled:	
Idiotic or insane:	

Household Members:

Name	Age
George Patterson	26
Rachel Patterson	24
Sallie Patterson	3
James O. Patterson	10m
Minnie H. Patterson	10m

Source Citation: Year: *1880*; Census Place: *Center, Oconee, South Carolina*; Roll *1236*; Family History Film: *1255236*; Page: *313B*; Enumeration District: *121*; .

Source Information:
Ancestry.com and The Church of Jesus Christ of Latter-day Saints. *1880 United States Federal Census* [database on-line]. Provo, UT, USA: Ancestry.com Operations Inc, 2010. 1880 U.S. Census Index provided by The Church of Jesus Christ of Latter-day Saints © Copyright 1999 Intellectual Reserve, Inc. All rights reserved. All use is subject to the limited use license and other terms and conditions applicable to this site.

Original data: Tenth Census of the United States, 1880. (NARA microfilm publication T9, 1,454 rolls). Records of the Bureau of the Census, Record Group 29. National Archives, Washington, D.C.

Description:
This database is an index to 50 million individuals enumerated in the 1880 United States Federal Census. Census takers recorded many details including each person's name, address, occupation, relationship to the head of household, race, sex, age at last birthday, marital status, place of birth, parents' place of birth. Additionally, the names of those listed on the population schedule are linked to actual images of the 1880 Federal Census. Learn more...

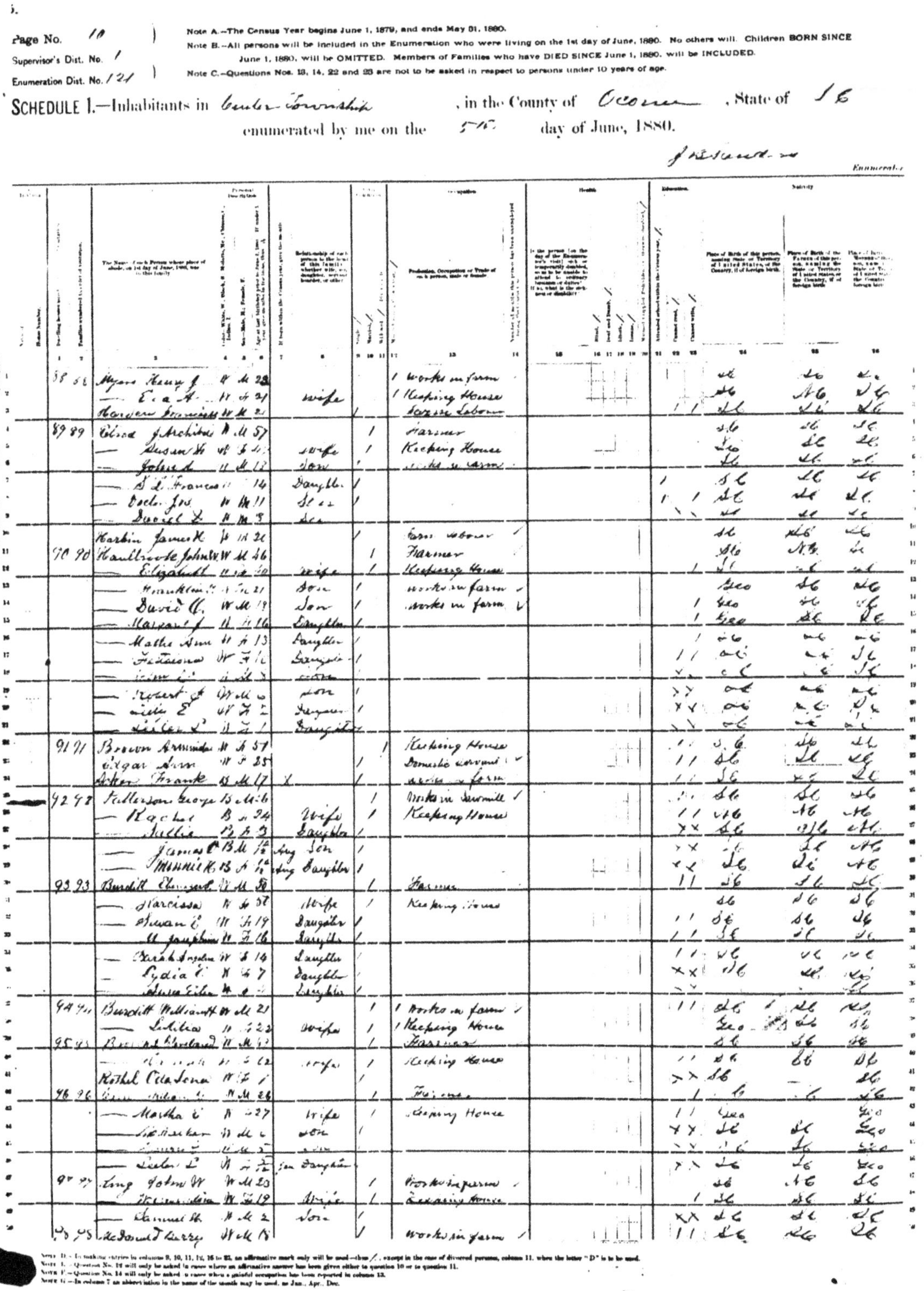

Page No. 10
Supervisor's Dist. No. 1
Enumeration Dist. No. 121

Note A.—The Census Year begins June 1, 1879, and ends May 31, 1880.
Note B.—All persons will be included in the Enumeration who were living on the 1st day of June, 1880. No others will. Children BORN SINCE June 1, 1880, will be OMITTED. Members of Families who have DIED SINCE June 1, 1880, will be INCLUDED.
Note C.—Questions Nos. 13, 14, 22 and 23 are not to be asked in respect to persons under 10 years of age.

SCHEDULE 1.—Inhabitants in Center Township, in the County of Oconee, State of SC enumerated by me on the 5th day of June, 1880.

[illegible], Enumerator.

Line	Dwelling	Family	Name	Color	Sex	Age	Relationship	Occupation	Birthplace	Father's birthplace	Mother's birthplace
1	88	88	Myers Henry J.	W	M	23		works in farm	SC	SC	SC
2			— Eva A.	W	F	21	wife	Keeping House	SC	NC	SC
3			Harden Francis	W	M	21		Farm Labor	SC	SC	SC
4	89	89	Elrod J. Archibald	W	M	57		Farmer	SC	SC	SC
5			— Susan H.	W	F	4[illegible]	wife	Keeping House	Geo	SC	SC
6			— John L.	W	M	18	Son	works in farm	SC	SC	SC
7			— S. L. Frances	W		14	Daughter		SC	SC	SC
8			— Doctor Jos.	W	M	11	Son		SC	SC	SC
9			— David L.	W	M	8	Son		SC	SC	SC
10			Harbin James H.	W	M	21		Farm labor	SC	SC	SC
11	90	90	Haulbrook John W.	W	M	46		Farmer	SC	NC	SC
12			— Elizabeth	W	F	40	wife	Keeping House	SC	SC	SC
13			— Franklin	W	M	21	Son	works in farm	Geo	SC	SC
14			— David C.	W	M	19	Son	works in farm	Geo	SC	SC
15			— Margaret J.	W	F	16	Daughter		Geo	SC	SC
16			— Mattie Ann	W	F	13	Daughter		SC	SC	SC
17			— Fetaiona	W	F	1[illegible]	Daughter		SC	SC	SC
18			— [illegible]	W	M	[illegible]	Son		SC	SC	SC
19			— Robert J.	W	M	[illegible]	Son		SC	SC	SC
20			— [illegible] E.	W	F	[illegible]	Daughter		SC	SC	SC
21			— Lilia L.	W	F	1	Daughter		SC	SC	SC
22	91	91	Brown Arminda	W	F	57		Keeping House	SC	SC	SC
23			Edgar Ann	W	F	25		Domestic servant	SC	SC	SC
24			Acker Frank	B	M	17		works in farm	SC	SC	SC
25	92	92	Patterson George	B	M	[illegible]		Works in Sawmill	SC	SC	SC
26			— Rachel	B	F	24	Wife	Keeping House	SC	SC	SC
27			— Nellie	B	F	3	Daughter		SC	SC	SC
28			— James C.	B	M	½	Aug Son		SC	SC	SC
29			— Minnie K.	B	F	½	Aug Daughter		SC	SC	SC
30	93	93	Burdett Ebenezer	W	M	5[illegible]		Farmer	SC	SC	SC
31			— Narcissa	W	F	58	Wife	Keeping House	SC	SC	SC
32			— Susan E.	W	F	19	Daughter		SC	SC	SC
33			— M. Josephine	W	F	16	Daughter		SC	SC	SC
34			— Sarah Angeline	W	F	14	Daughter		SC	SC	SC
35			— Lydia E.	W	F	7	Daughter		SC	SC	SC
36			— Avice Ellen	W	F	[illegible]	Daughter				SC
37	94	94	Burdett William H.	W	M	21		Works in farm	SC	SC	SC
38			— Letitia	W	F	22	Wife	Keeping House	Geo	SC	SC
39	95	95	Br[illegible] Cleveland	W	M	[illegible]		Farmer	SC	SC	SC
40			— [illegible]	W	F	62	wife	Keeping House	SC	SC	SC
41			Rothel Cila Lona	W	F	1			SC	—	SC
42	96	96	[illegible]	W	M	26		Farmer	SC	SC	SC
43			— Martha E.	W	F	27	Wife	Keeping House	Geo		Geo
44			— [illegible]	W	M	6	Son		SC	SC	Geo
45			— [illegible]	W	M	[illegible]	Son		SC	SC	Geo
46			— Lula L.	W	F	[illegible]	Jan Daughter		SC	SC	Geo
47	97	97	Long John W.	W	M	23		Works in farm	SC	SC	SC
48			— [illegible]	W	F	19	Wife	Keeping House	SC	SC	SC
49			— Samuel H.	W	M	2	Son		SC	SC	SC
50	98	98	McDonald Berry	W	M	18		works in farm	SC	SC	SC

Note D.—In making entries in columns 9, 10, 11, 12, 16 to 23, an affirmative mark only will be used—thus /, except in the case of divorced persons, column 11, when the letter "D" is to be used.
Note E.—Question No. 12 will only be asked in cases where an affirmative answer has been given either to question 10 or to question 11.
Note F.—Question No. 14 will only be asked in cases where a gainful occupation has been reported in column 13.
Note G.—In column 7 an abbreviation in the name of the month may be used, as Jan., Apr., Dec.

PATTERSON FAMILY

State Democrats organized parades and rallies in every county of South Carolina. Many of the participants were armed and mounted; all wore red. Mounted men gave an impression of greater numbers. When Wade Hampton and other Democrats spoke, the Red Shirts would respond enthusiastically, shouting the campaign slogan, "Hurrah for Hampton." This created a massive spectacle that united and motivated those present.

Red Shirts sought to intimidate both white and black watchers into voting for the Democrats or even not at all. The Red Shirts and similar groups were especially active in those few states with an African-American majority. They broke up Republican meetings, disrupted their organizing, and intimidated black voters at the polls. Many freedmen stopped voting from fear, and others voted for Democrats under pressure. The Red Shirts did not hesitate to use violence, nor did the other private militia groups. In the Piedmont counties of Aiken, Edgefield, and Barnwell, freedmen who voted were driven from their homes and whipped, while some of their leaders were murdered. During the 1876 presidential election, Democrats in Edgefield and Laurens counties voted "early and often", while freedmen were barred from the polls:

State Democrats organized parades and rallies in every county of South Carolina. Many ofthe participants were armed and mounted; all wore red. Mounted men gave an impression of greater numbers. When Wade Hampton and other Democrats spoke, the Red Shirts would respond enthusiastically, shouting the campaign slogan, "Hurrah for Hampton." This created a massive spectacle that united and motivated those present.

Red Shirts sought to intimidate both white and black watchers into voting for the Democrats or even not at all. The Red Shirts and similar groups were especially active in those few states with an African-American majority. They broke up Republican meetings, disrupted their organizing, and intimidated black voters at the polls. Many freedmen stopped voting from fear, and others voted for Democrats under pressure. The Red Shirts did not hesitate to use violence, nor did the other private militia groups. Inthe Piedmont counties of Aiken, Edgefield, and Barnwell, freedmen who voted were driven from their homes and whipped, while some oftheir leaders were murdered. During the 1876 presidential election, Democrats inEdgefield and Laurens counties voted "early and often", while freedmen were barred

Armed and mounted Red Shirts accompanied Hampton on his tour of the state. They attended Republican meetings and would demand equal time, but they usually only stood in silence. At times, Red Shirts would hold a barbecue nearby to lure Republicans and try to convince them to vote for the Democratic ticket.
Hampton positioned himself as a statesman, promising support for education and offering protection from violence that Governor Daniel Henry Chamberlain did not seem able to provide. Few freedmen voted for Ha and most remained loyal to the Republican Party of Abraham Lincoln. The 1876 campaign was the "most tumultuous in South Carolina's history." "An anti-Reconstruction historian later estimated that 150Negroes were murdered in South Carolina during the campaign."

After the election on November 7, a protracted dispute between Chamberlain and Hampton ensued as both claimed victory. Because of the massive election fraud, Edmund William McGregor Mackey, a Republican member of the South Carolina House of Representatives, called upon the "Hunkidori Club" from Charleston to eject Democratic members from Edgefield and Laurens counties from the House. Word spread through the state. By December 3, approximately 5,000 Red Shirts assembled at the State House to defend the Democrats. Hampton appealed for calm and the Red Shirts dispersed.

As a result of a national political compromise, President Rutherford B. Hayes ordered the removal of the Union Army from the state on April 3, 1877. The white Democrats completed their political takeover of South Carolina. In the gubernatorial election of 1878, the Red Shirts made a nominal appearance as Hampton was re-elected without opposition.

Future South Carolina Democratic politicians such as Benjamin Tillman and Ellison D. Smith, proudly claimed their association with the Red Shirts as a *bonafide* for white supremacy.

ancestry

Name:	**Jos Crews**
Age in 1860:	37
Birth Year:	abt 1823
Birthplace:	North Carolina
Home in 1860:	Laurens, Laurens, South Carolina
Gender:	Male
Post Office:	Laurens
Value of real estate:	View image

Household Members:

Name	Age
Adam Crews	9
Agustus Crews	5
Hester Crews	2
No Name Crews	1/12
Jos Crews	37
Malinda Crews	26

Source Citation: Year: *1860*; Census Place: *Laurens, Laurens, South Carolina*; Roll *M653_1222*; Page: *220*; Image: *6*; Family History Library Film: *805222*.

Source Information:
Ancestry.com. *1860 United States Federal Census* [database on-line]. Provo, UT, USA: Ancestry.com Operations, Inc., 2009. Images reproduced by FamilySearch.

Original data: 1860 U.S. census, population schedule. NARA microfilm publication M653, 1,438 rolls. Washington, D.C.: National Archives and Records Administration, n.d.

Description:
This database is an index to individuals enumerated in the 1860 United States Federal Census, the Eighth Census of the United States. Census takers recorded many details including each person's name, age as of the census day, sex, color; birthplace, occupation of males over age fifteen, and more. No relationships were shown between members of a household. Additionally, the names of those listed on the population schedule are linked to actual images of the 1860 Federal Census. Learn more...

1880 United States Federal Census- Ancestry.com

Name:	Malinda Crews [Malinda Dial]
Home in 1880:	Laurens, Laurens, South Carolina
Age:	45
Estimated birth year:	abt 1835
Birthplace:	South Carolina
Relation to Head of Household:	Self *(Head)*
Father's birthplace:	South Carolina
Mother's Name:	Manerva Dial
Mother's birthplace:	South Carolina
Neighbors:	View others on page
Occupation:	Farmer
Marital Status:	Widowed
Race:	White
Gender:	Female
Blind:	
Deaf and dumb:	

Household

	Name	Age
Members:	Malinda Crews	45
	Hettie M.Crews	20
	Sidney M. Crews	19
	Rose M. Crews	15
	Horrace J. Crews	12
	Louies Crews	9
	Joseph W. Crews	4
	Manerva Dial	65
	Geo. G. Winslow	63
	Lavinda Bolt	40
	Celey Bolt	10
	Nancy Williams	80

Source Citation Year: *1880*; Census Place: *Laurens, Lourens, South Carolina*; Roll J2:J3: Family History Film: *1:255233*: Pa ; : *12C*; Enumration Di.strict: *99*.

http://search.ancestry.cornlcgi-bin/sse.dll?db=1880usfedcen&indiv=try&h=4248163 7 10118/2010

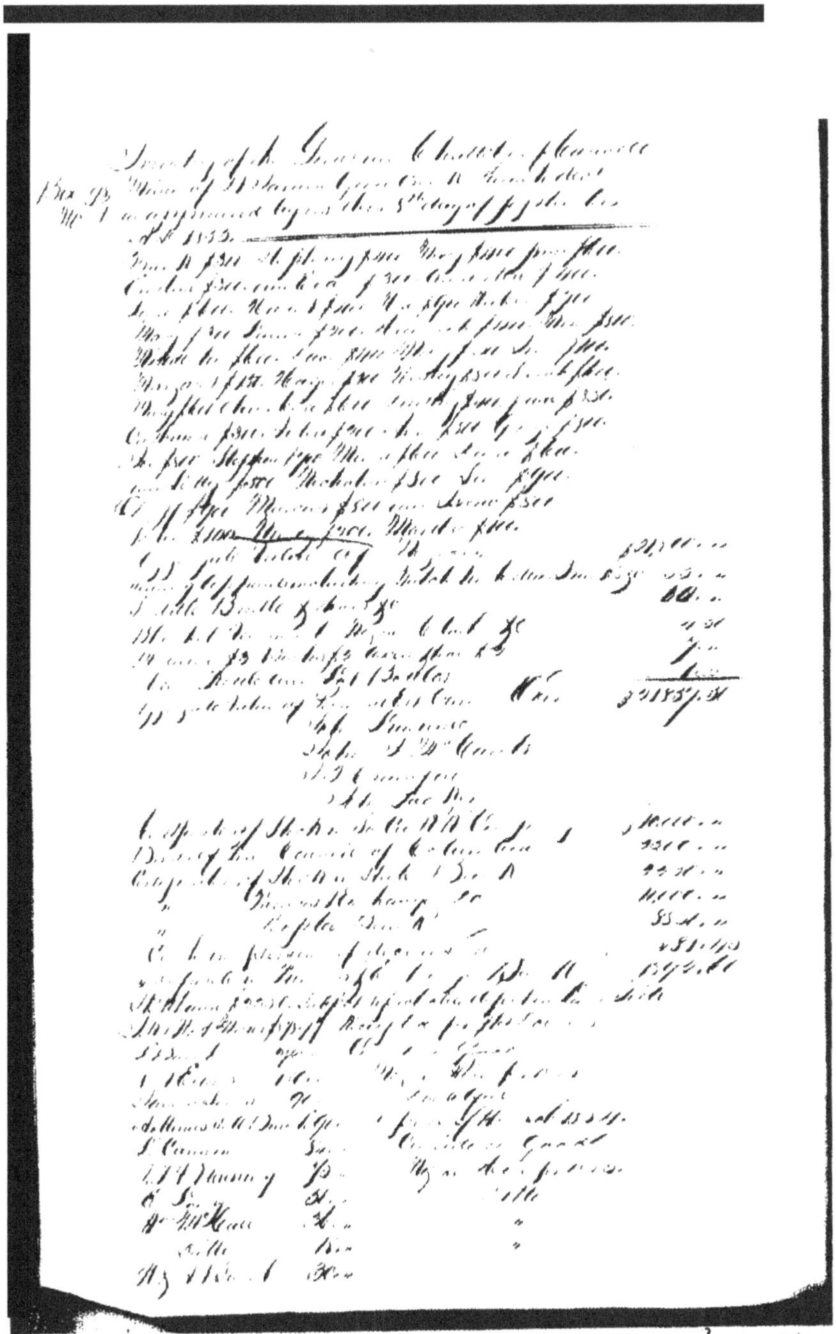

Goose Creek

John J. McCants 1855

S. W. Haven $19.20 Bad
A. Stanley 30.00 Negro Since present years
W. B. Hadley 25.00 Bad
In addition there is due the following accounts
J. W. Steele $23.43 Twenty three 43/100 Dollars S. Goodland
$95 Ninety five Dollars [illegible] 51.90 Fifty one
90/100 Dollars E. [illegible] $23.93 Twenty three
93/100 Dollars S. W. [illegible] $20 Twenty Dollars
J. P. West $10 Ten Dollars

Filed 18 Sept 1855. John Nelson Admr

Bk 94 No 3 Inventory of the Estate of Elias Taylor deceased

Claim on the United States Government for extra pay due for services rendered on Board the U. S. Ship Columbus on the Coast of California in 1847 Mary S. Taylor

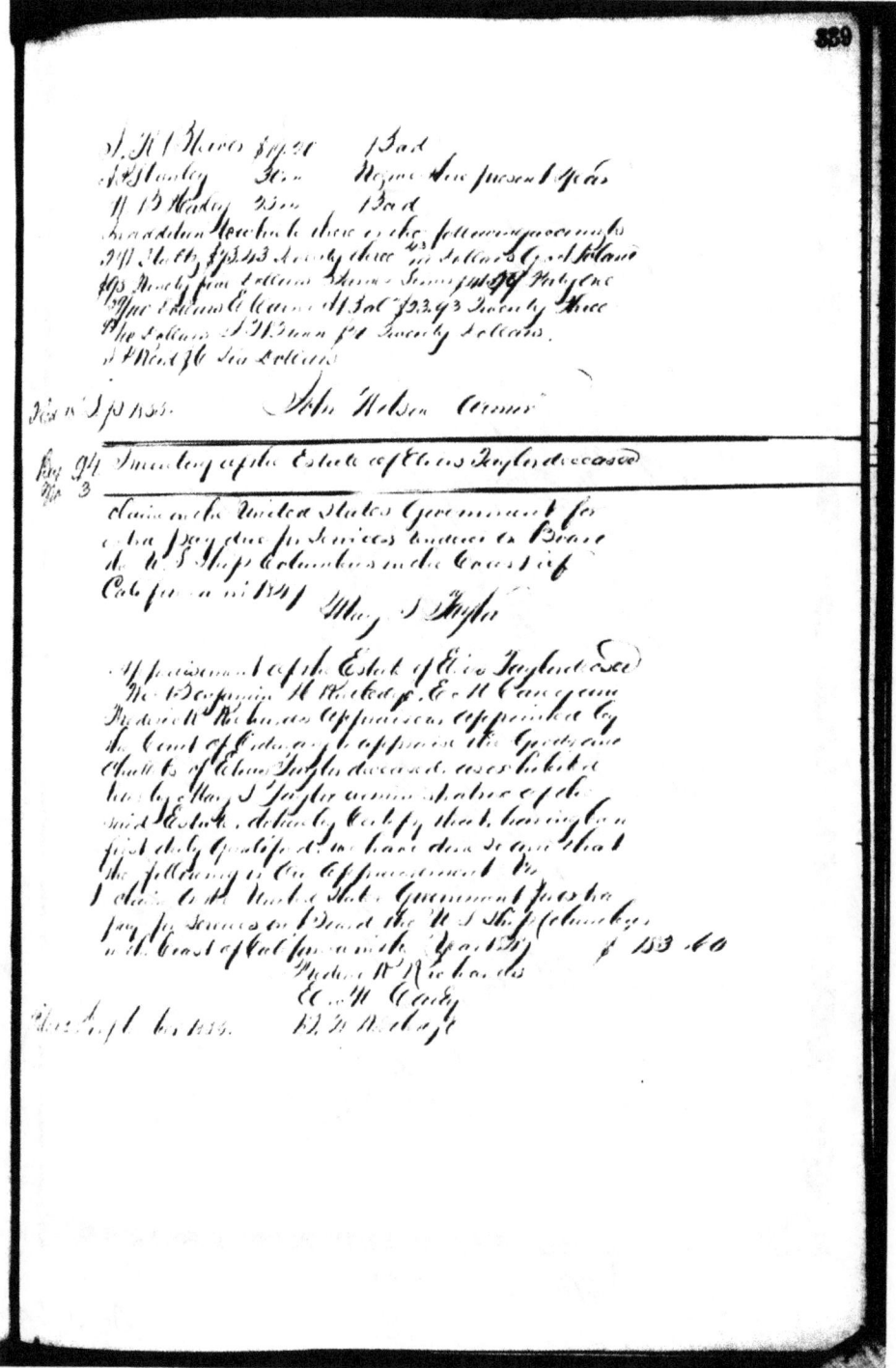

Appraisement of the Estate of Elias Taylor Decd
We Benjamin H. Hadley, E. H. Cary and Frederick Richards Appraisers appointed by the Court of Ordinary to appraise the Goods and Chattels of Elias Taylor deceased exhibited here by Mary S. Taylor administratrix of the said Estate, do hereby Certify that having been first duly Qualified we have done so and that the following is our Appraisement viz
1 claim on the United States Government for extra pay for services on Board the U. S. Ship Columbus on the Coast of California in the year 1847 $ 153.60
Frederick Richards
E. H. Cary
Filed Sept 6th 1855. B. H. Hadley

John Conner [illegible]

The appraise Bill of the Estate of John Conner Decd [illegible] fellow Bill appraised at 1000 [illegible] [illegible] [illegible] [illegible] George 250 [illegible] [illegible] [illegible] cattle [illegible] [illegible] Wagon [illegible] [illegible] [illegible] [illegible] 20 [illegible] [illegible] plantation tools [illegible] [illegible] [illegible] [illegible] [illegible] [illegible] Clock [illegible] [illegible] [illegible] furniture [illegible] [illegible] furniture [illegible] 2 [illegible] [illegible] 2 [illegible] 2. $2272.00

This is to certify that we the appraisers called on by [illegible] [illegible] to appraise the Goods & chattels of John Conner Decd do certify that the above is a true statement of the above Estate
[illegible] Jones [illegible] Davis
[illegible] McKee

Wm G McCants

Inventory and appraisement of the Goods and chattels of the Estate of William [illegible] late [illegible] Christ Church Parish [illegible] [illegible] by Archibald Campbell [illegible] John Porteous [illegible] qualified [illegible] of said estate — At the plantation called [illegible]

Negro Slaves

3	[illegible] [illegible] Nancy 500 [illegible]	1700.00
4	[illegible] [illegible] Jimmie 600 [illegible] [illegible] 200	[illegible]
[illegible]	[illegible] 1000 [illegible] 800	1800
[illegible]	[illegible] [illegible] [illegible] 400 [illegible] 400	[illegible]
4	[illegible] 800 [illegible] 400 William 3 [illegible] 100	1500
2	[illegible] 500 Paul 1000	1500
	[illegible] 500 Sally 800 [illegible]	
4	Little [illegible] 600	2300
[illegible]	[illegible] 500 Betty 800 Robert 600 [illegible]	[illegible]
2	[illegible] 800 Sarah 800	1600
2	[illegible] 500 [illegible] 20	[illegible]
	[illegible] 700 [illegible] 300 Jacob [illegible]	
	Jeffery [illegible] Isaac 700 [illegible] [illegible]	
7	[illegible] 300	4500
4	[illegible] 150 Philis 500 [illegible] [illegible]	[illegible]
1	Pally 800	800
1	[illegible] 600	600
1	Neptune 200	200
4	Robert 600 Richard [illegible] [illegible] 800 [illegible]	2800
2	[illegible] 100 Charlotte [illegible]	[illegible]

Continued

[illegible] Jerry 600	$ 600	
[illegible] Eliza 600	600	
[illegible] Nancy 600	600	
[illegible] Children [illegible]	[illegible]	
Nancy [illegible] Frederick 1000 [illegible] in the Will [illegible]	1800	
[illegible] 700 [illegible] in the Will [illegible]	1400	
[illegible] Negro Slaves		$34455
The following [illegible] property of the Estate [illegible] possession of [illegible] the Will [illegible] children [illegible]		
William [illegible] Kate 1000 [illegible] Estate for the present		

Cattle Horses &c.

18 Head of stock [illegible]	[illegible]	
[illegible] Head of cattle [illegible]	150	
3. Head of sheep [illegible]	[illegible]	
[illegible] Mules [illegible]	[illegible]	
[illegible] of carriage Horses	400	
[illegible]	50	
Gray Horse [illegible]	[illegible]	
[illegible]	10	
2 Wagons	[illegible]	
[illegible] Carts	[illegible]	
About 3[illegible] Bags Cotton	3000	
500 Bush Corn at 75	362[illegible]	
Lot of Fodder [illegible]	100	
[illegible] cotton [illegible]	[illegible]	
[illegible] $25 [illegible]	75	5627

Furniture at [illegible] & at [illegible] Pleasant Residence

10 Bedsteads [illegible] 15 [illegible]	400	
1 Feather Bed 25 15 Pillows $15	40	
8 Bolsters $16 [illegible] at 15	3[illegible]	
2 P[illegible] Card Tables 20 1 Sofa 12	32	
2 [illegible] Tables [illegible] $20	20	
Clock 13 2 Sideboards 2[illegible]	23	
[illegible]	34	
2 doz [illegible] 1 doz [illegible] Spoons [illegible]	130	
11 Tea Spoons 11 18 Silver Forks [illegible]	[illegible]	
[illegible] Blankets bed clothing 1 Table [illegible] 40 [illegible]	150	
[illegible] Fishing [illegible] Gun 50	60	984
		$43.068

(carried over)

A true and perfect Inventory and Appraisement of the Estate of Mrs Eliza McCants deceased as produced to us by Archd Campbell Executor in Colleton Dist.
March 13th 1843 Recd July 8th 1843
Negro Slaves James $500 Toby $250 Bob $400
Grace $400 Nancy $175 Dorcas $400 Violet $100 } $2225.
Given under our hands at Round O in St Bartholo-
:mews Parish March 13th 1843
Fredk Fraser R. Fishburne Josiah Beck

©GraphicMaps.com
Senegal River
GUINEA
LOW / HILLS
MTS.
Senegal
Mali
CLICK HERE FOR LARGER MAP
Gambia River
Gambia
100 mi
100 km
Bamako
Guinea Bissau
Kifaya
Mali
Bafing River
Konkoure River
Labe
Siguiri
Boke
Niger River
Kamsar
Niger Nat. Park
Cape Verga
Kankan
Boffa
Fria
Kindia
Guinea
Conakry
Niger River (source)
Milo River
Freetown
Sierra Leone
Macenta
Atlantic Ocean
Nzerekore
Mt. Nimba
Liberia
Cote D'Ivoire
worldatlas
Monrovia

SLAVE NARRATIVES
Impressive colletion of interviews of former slaves in the United States

All *Slave Narratives* results for *George Fleming*

Searching for...

Name: George Fleming
Birth: South Carolina, USA
Death: South Carolina, USA

Edit Search

or Start a new search

Narrow by Collection

All Categories
Stories, Memories & Histories
Family Histories, Journals & Biographies
Slave Narratives

ABOUT *SLAVE NARRATIVES*

Impressive collection of interviews of former slaves in the United States

Learn more about this database

Hot Keys

New search
Refine search
Preview current record
Highlight next record
Highlight previous record

Matches **53,103-53,112** of **80,836**

State: South Carolina

Interviewee: Fleming, George

Fleming, George

State: South Carolina

Interviewee: Fleming, George

George Fleming and his wife, Elizabeth, live in a small two-room cottage at 349 Highland Street in Spartanburg, S. C. Their humble abode is typical of the average negro dwelling in this city. It is furnished with only the bare necessities compatible with comfortable living; but to George and Elizabeth it holds the same warmth and feeling of security which their idea of a home depicts. George has a keen memory and he talked freely of slavery.

State: South Carolina

Interviewee: Fleming, George

"I was born in 1854 in de month of August. I disremembers what dat pension lady said was de day. She de one dat found out all about it. I 'clar dat was de biggest plantation whar I was born dat I is ever seed or heard tell of. Lawd a-mercy! Ain't no telling how many acres in dat place, but dar was jes' miles and miles of it. It was in Laurens County, not fur (far) frum de town of Laurens. I 'longed to Marse Sam Fleming. Lawd chile, dat's de best white man what ever breathed de good air. I still goes to see whar he buried every time I gits a chance to venture t'wards Laurens. As old as I is, I still draps a tear when I sees his grave, fer he sho was good to me and all his other niggers.

State: South Carolina

Interviewee: Fleming, George

"Marse Sam's boys, Lyntt and Frank, sho was tigers, but cose dey wasn't mean tigers. Dey had real long beards. Marse Lyntt was my young marster, and he de bestest man I ever know'd, 'cepting his daddy. He allus doing something to have fun outen us lil' niggers; but us didn't mind, 'cause we got fun outen it, too. I 'member how he used to sot us in de hog pens, but we wasn't scared as we 'lowed we was.

State: South Carolina

Interviewee: Fleming, George

"My pa named Bill, He was stole frum Virginia. I don't know how Marse got him. Sometimes dey would buy 'em and agin dey would steal 'em, sort of like stealing a dog. Ma, her name Hannah. Dey got married on de plantation. Atter pa got kil't, ma married a man called Charles. I only has one whole sister living, and she name Jennie. Viney and Millie be dead. My brother, Richard, he dead, too. My half-sister, Sallie Ann, she stay in Jacksonville, or some foreign state.

State: South Carolina

Interviewee: Fleming, George

"Mercy on us, dem was de happy days; dey was heavenly days 'sides what we 'speriences now. Us lil' kids played lots of games den, some of dem like what dey plays now, but we had a better time. Befo' we was big enough to work, 'cept tote water and de like of dat, we played sech things as marbles. We had purty red and blue marbles dat Marse Lyntt brung frum de store. Sometimes we wrestle, too, and old Marse laugh till his fat belly shake all over when he see de lil' nigger's head buried in de white sand. Sometimes we play 'warm jacket'. Dat was worked by each one gitting a brush frum a tree or bush and frailing de other'un till it got too hot fer him.

State: South Carolina

Interviewee: Fleming, George

"De older boys and gals had big frolics, 'specially in de fall of de year. Sometimes dey be on our plantation, and agin dey be on neighboring ones. When dey have 'em close home, some of us lil' niggers would slip off and git in de corner or up in de loft of de house and spy on 'em. Dey cotch us sometimes and thrash us out. One game dey played was 'please and displease'. When de gal say, 'What it take to please you?' de boy say, 'A kiss frum dat purty gal over dar'. Yes, dey played 'hack-back', too. Dat's when dey faced each other and trotted back and forth. Lawd, dey sho had some awful times dancing and cutting jigs. Twan't much drinking, 'cepting on de side.

State: South Carolina

Interviewee: Fleming, George

"White ladies didn't go to de frolics, but some of de white men did. De partollers was allus around to see dat everybody had passes, and if dey didn't have 'em dey was run back home. Sometimes de overseer was dar, too. Lawd, dey sho did kick up de dust at dem frolics. De music was mostly made by fiddles, and sometimes dey had quill blowers. De quills was made frum cane, same as de spindles was but dey was cut longer and was different sizes. All de quills was put in a rack and you could blow any note you wanted to off of dem. Boy, I sho could blow you out of dar wid a rack of quills. I was de best quill blower dat ever put one in a man's mouth. I could make a man put his fiddle up; hit you so hard wid Dixieland dat I knock you off de seat. Gals wouldn't look at nobody else when I start blowing de quills.

State: South Carolina

Interviewee: Fleming, George

"Dar was also heaps and lots of other big affairs 'sides de frolics. De cornshuckings -- Lawd a-mercy, you ain't seen nothing. Niggers frum all over de place shucking corn and somebody setting on one of de big piles calling de cornshucking song, jes' like dey do in de square dance. Dat kept 'em happy --- everybody jine in de chorus. A jug of liquor sot at de bottom of de pile; everybody try to be first to get to de liquor. Lawd, dey holler and take on something awful when dey get to de bottom. White folks have big supper ready; liquor, brandy and everything. Dem was de times; pick up somebody and kivver 'em up wid de shucks. Had cotton pickings, too. Dat work not so fast but we had good times. Sometimes dey be on our plantation; den we sometimes go to other places.

State: South Carolina

Interviewee: Fleming, George

"Didn't need no passes when a bunch of slaves went to other plantations to dem big gatherings. 'Rangements was already made so de patrollers wouldn't bother nobody. Dat policy didn't hold fer de frolics, though. Sho had to have a pass frum de marse if you went.

All *Slave Narratives* results for *George Fleming*

Searching for...

Name: George Fleming
Birth: South Carolina, USA
Death: South Carolina, USA

Edit Search
or Start a new search

Narrow by Collection

All Categories
Stories, Memories & Histories
Family Histories, Journals & Biographies
Slave Narratives

ABOUT *SLAVE NARRATIVES*

Impressive collection of interviews of former slaves in the United States

Learn more about this database

Hot Keys

New search
Refine search
Preview current record
Highlight next record
Highlight previous record

Matches **53,113-53,122** of **80,836**

State: South Carolina

Interviewee: Fleming, George

"On de plantation we lived jes' like a great big family wid Marse de daddy of 'em all. Cose he had overseers to watch atter de work and keep things straight. He allus kept more dan 200 head of slaves. De quarters was made up of lots of cabins, some wid one room, some wid two or three. Dat's 'cording to how big de family was. Dey wasn't built in rows, but scattered about over de plantation. Some of de cabins was made of logs and some wid planks, but all was warm and comfortable. Dey had all kinds of chimneys, too. Some brick, some rock and some de old stick and mud kind. Dey all had big fireplaces. Dat was whar us done de cooking. Hitches (hooks) was on de sides of de fireplace whar big iron pots hung to bile and cook in. We had pans and leads (lids) and things to bake in, too. Yes Lawd, dem was de days, fer we sho had plenty to eat -- everything we wanted.

State: South Carolina

Interviewee: Fleming, George

"All de things we had in de house was home-made, but we sho had good beds. Dey made wid boards, and 'stead of slats, ropes was stretched twixt de sides real tight by slipping dem through holes and making knots in de ends. Over dese we laid bags; den feather or straw ticks. We had plenty kivvers to keep us warm. We had shelves and hooks to put our clothes on. We had benches and tables made wid smooth boards. Missus Harriet, dat Marse Sam's wife, she give us a looking-glass so we could see how to fix up. Lawd a-mercy, Missus Harriet was one fine woman. She allus looked atter us to see dat we didn't suffer fer nothing.

State: South Carolina

Interviewee: Fleming, George

"Some of de women dat didn't have a passel of lil' brats was 'signed to de job of cooking fer de field hands. Some of 'em come home to eat, but mostly dey stayed in de fields. De dinner horn blow'd 'zactly at 12 o'clock and dey know'd it was time fer grub. Everybody drapped what dey was doing and compiled demselves in groups. Dey could see de buckets coming over de hill. Dar was more dan one group, fer de fields was so big dat dey couldn't all come to one place. Cose all dat was planned out by de overseers. Had lots of overseers and dey had certain groups to look out fer.

State: South Carolina

Interviewee: Fleming, George

"Most of de food was brung to de fields in buckets, but sometimes de beans and de like of dat come in de same pots dey was cooked in. It took two big niggers to tote de big pots. Dar was no want of food fer de hands. Marse know'd if dey worked dey had to eat. Dey had collards, turnips and other good vegetables wid cornbread. Chunks of meat was wid de greens, too, and us had lots of buttermilk.

State: South Carolina

Interviewee: Fleming, George

"Women worked in de field same as de men. Some of dem plowed jes' like de men and boys. Couldn't tell 'em apart in de field, as dey wore pantelets or breeches. Dey tied strings 'round de bottom of de legs so de loose dirt wouldn't git in deir shoes. De horn blow'd to start work and to quit. In de morning when de signal blow'd, dey all tried to see who could git to de field first. Dey had a good time and dey liked to do deir work. Us didn't pay much mind to de clock. We worked frum sun to sun. All de slaves had to keep on de job, but dey didn't have to work so hard. Marse allus said dey could do better and last longer by keeping 'em steady and not overworking 'em.

State: South Carolina

Interviewee: Fleming, George

"Dar was all kinds of work 'sides de field work dat went on all de time. Everybody had de work dat he could do de best. My daddy worked wid leather. He was de best harness maker on de place, and he could make shoes. Dey had a place whar dey tanned cow-hides. Dat was called de tannos. Dey didn't do much spinning and weaving in de home quarters; most of it was done in one special place Marse had made fer dat purpose. Some of de slaves didn't do nothing but spin and weave, and dey sho was good at it, too. Dey was trained up jes' fer dat particular work.

State: South Carolina

Interviewee: Fleming, George

"I don't know how many spinning wheels and looms and dem things Marse had, but he sho had lots of 'em. Dat business making cloth had lots to it and I don't know much 'bout it, but it was sort of dis way. Dey picked de seeds out of de cotton; den put de cotton in piles and carded it. Dey kept brushing it over and over on de cards till it was in lil' rolls. It was den ready fer de spinning wheels whar it was spun in thread. Dis was called de filling. I don't know much 'bout de warp, dat is de part dat run long ways.

State: South Carolina

Interviewee: Fleming, George

"Dem spinning wheels sho did go on de fly. Dey connected up wid de spindle and it go lots faster dan de wheel. Dey hold one end of de cotton roll wid de hand and 'tach de other to de spindle. It keep drawing and twisting de roll till it make a small thread. Sometimes dey would run de thread frum de spindle to a cornshuck or anything dat would serve de purpose. Dat was called de broach. Some of dem didn't go any further dan dat, dey had to make sech and sech broaches a day. Dis was deir task. Dat's de reason some of dem had to work atter dark, dat is, if dey didn't git de task done befo' dat.

State: South Carolina

Interviewee: Fleming, George

"Dey run de thread off de broach on to reels, and some of it was dyed on de reels. Dey made deir own dyes, too. Some of it was made frum copperas, and some frum barks and berries. Atter while, de thread was put back on de spinning wheel and wound on lil' old cane quills. It was den ready fer de looms. Don't know nothing, de looms - boom! boom! sho could travel. Dey put de quills, atter de thread was wound on dem, in de shettle and knocked it back and forth twixt de long threads what was on de beams. Can't see de thread fly out of dat shettle it come so fast. Dey sho could sheckle it through dar. Dey peddled dem looms, zip! zap! making de thread rise and drap while de shettle zoom twixt it. Hear dem looms booming all day long 'round de weaving shop. De weaving and spinning was done in de same place.

State: South Carolina

Interviewee: Fleming, George

"Overseers lived on de plantation. No, dey wasn't poor whites. All Marse Sam's overseers was good men. Dey lived wid deir families, and Marse's folks 'sociated wid dem, too. Dey had good houses to live in. Dey built better dan ours was. Marse didn't 'low dem to whip de slaves, but dey made us keep straight. If any whipping had to be done, Marse done it, but he didn't have to do much. He didn't hurt 'em bad, den, jes' git a big hickry and lay on a few. He would say if dat nigger didn't walk de chalk, he would put him on de block and settle him. Dat was usually enough, 'cause Marse mean't dat thing and all de niggers know'd it.

All *Slave Narratives* results for *George Fleming*

Searching for...

Name: George Fleming
Birth: South Carolina, USA
Death: South Carolina, USA

Edit Search
or Start a new search

Narrow by Collection

All Categories
Stories, Memories & Histories
Family Histories, Journals & Biographies
Slave Narratives

ABOUT *SLAVE NARRATIVES*

Impressive collection of interviews of former slaves in the United States

Learn more about this database

Hot Keys

New search
Refine search
Preview current record
Highlight next record
Highlight previous record

Matches **53,123-53,132** of **80,836**

State: South Carolina

Interviewee: Fleming, George

"Jes' one or two of Marse Sam's slaves ever run away, but lots of other niggers did. Some of dem try to go to de North, but mostly dey come polling back by demselves when dey git hungry. If dey didn't come back purty soon, deir marse sont out to look fer 'em. Lawd, I heard de nigger hounds yelping befo' day many times. Dat was de bloodhounds dey sicked on de runaway niggers, and dey sho run 'em back home. When dey hear de hounds dey was glad to git home.

State: South Carolina

Interviewee: Fleming, George

"De patrollers would go out and look fer de niggers. Dey almost skin 'em alive if dey cotch 'em befo' dey git home. Patrollers was made up of jes' anybody dat wanted to jine 'em, poor white trash and all. One thing dey sho couldn't do, and dat was tech a nigger atter he done got on his marse's grounds. Dey almost got pa one time, but he saved his hide by falling over a rail fence jes' befo' dey cotched him. All de plantation owners, dey pay so much to de patrollers to be on de look-out fer de slaves, and dat's de way dey kept so many frum running away.

State: South Carolina

Interviewee: Fleming, George

"Some men, like old Joe Crews, was reg'lar nigger traders. Dey bought niggers, stole 'em frum Virginia and places and drove 'em through de country like a bunch of hogs. Dey come in great gangs. In town dey have big nigger sellings, and all de marsters frum all over de countryside be dar to bid on 'em. Dey put 'em up on de block and holler 'bout dis and dat dey could do and how strong dey was. 'Six hundred --- Yip, Yip, make it six-fifty' I heard 'em call many times when I be dar wid Marse. Some of dem throw a thousand dollars quick as dey would ten at a purty gal. Some traders stop a drove of niggers at de plantation and swap or sell some. Dey didn't call dat putting 'em on de block like when day had de big selling.

State: South Carolina

Interviewee: Fleming, George

"Slaves started to work by de time dey was old enough to tote water and pick up chips to start fires wid. Some of dem started to work in de fields when dey about ten, but most of 'em was older. Lawd, Marse Sam must have had more dan a dozen house niggers. It took a lot of work to keep things in and 'round de house in good shape. Cose most of de slaves was jes' field hands, but some of dem was picked out fer special duties. Slaves didn't get any pay in money fer work, but Marse give 'em a lil' change sometimes.

State: South Carolina

Interviewee: Fleming, George

"Everybody have plenty to eat. Lots of times we had fish, rabbits, possums and stuff like dat; lots of fishing and hunting in dem days. Some slaves have lil' gardens of deir own, but most de vegetables come frum de big garden. Missus was in charge de big garden, but cose she didn't have to do no work. She sho seed atter us too. Even de poor white trash had plenty to eat back in dem times. Marse have a hundred head of hogs in de smokehouse at one time. Never seen so much pork in my life. We sho lived in fine fashion in hog killing time, cose de meats was cured and us had some all de year. Yes sir, Marse ration out everybody some every week. Watermelons grow awful big, some of 'em weigh a hundred pounds. Dey big striped ones, called 'rattlesnakes', so big you can't tote it no piece. All de baking and biling was done over de big fireplaces.

State: South Carolina

Interviewee: Fleming, George

"Didn't wear much clothes in summer 'cause we didn't need much, but all de grown niggers had shoes. Lawd, I wore many pair of Marse Lyntt's boots, I means sho 'nuff good boots. Marse had his own shoemakers, so twan't no use us gwine widout. Had better clothes fer Sunday. Most de washing was done on Saturday afternoons, and we be all setting purty fer Sunday. Cold weather we was dressed warm, and we had plenty bed kivvers, too. Cose all slaves didn't have it as good as Marse Sam's did. Lawd, I is seed lil' naked niggers setting on de rail fences like pa'cel of buzzards; but Marse Sam's niggers never had to go dat way.

State: South Carolina

Interviewee: Fleming, George

"Slaves didn't have no church or schools. Lots of dem went to de white folks' church, but Marse Sam didn't make his slaves go if dey didn't want to. Deir benches was on de sides and in de back of de church. All preachers was white men. Old preacher Moore sho was a humdinger, and a good one. He pizen deir minds wid Salvation, soak 'em in de oil of Holy Ghost and set 'em on fire. Lawd-a-me! When he got lit up all over till his eyes shine and sparkle, he sho could bring down de house. Twan't no seats in school fer de slaves, though. Some of de slick ones slipped around and larn't de letters.

State: South Carolina

Interviewee: Fleming, George

"When de slaves come from de field, deir day's work was done. Fact is, everybody's work was done 'cept maybe some of de spinners or weavers dat didn't quite finish deir task. Dey was de onliest ones dat had to ever work atter dark, and dat not often. Sometimes on Saturdays we didn't have to work a-tall, dat is in de fields, and sometimes we had to work till 12 o'clock. Lots of de men went fishing and hunting, and mostly de women washed. Saturday nights some groups would git together and sing. I can still hear dem old songs in my mind, but I doesn't recalls de words. Christmas sho was handsome time. Christmas and New Years we had a good time. Marse jes' sort of turned 'em loose. We got a lil' extra liquor and brandy on de holidays, but cose we had some all along enduring de whole entire year. Marse had three stills on de place and dar was plenty liquor but he didn't let anybody git drunk. He call de lil' niggers, too, sometimes and give 'em a drink, and he give 'em jelly biscuits. He call everybody up to de big house on Christmas and make a speech; den he give everybody some good brandy.

State: South Carolina

Interviewee: Fleming, George

"I doesn't recalls nothing 'bout no ghosts. Ain't nothing in dem things. Cose if you goes 'round de graveyards atter dark, you might see sech things, but I ain't gwine dar. Nigger come in once a-telling something 'bout a witch making a knot in his horse's tail, but I don't think dar was nothing to it.

State: South Carolina

Interviewee: Fleming, George

"When any of de slaves got sick, Marse took good care of 'em till dey got well. If dey bad sick he sont fer de doctor. Some of de women know'd how to bile up herbs and roots and make tea fer colds and fevers, but I don't know what kind dey used. When de chilluns was born, Marse seed to it dat de mammy was rightly took care of. He kept a old granny woman wid dem till dey got up and well.

All *Slave Narratives* results for *George Fleming*

Matches **53,133-53,142** of **80,836**

State: South Carolina

Interviewee: Fleming, George

"De slaves mostly got married in Marse Sam's back yard, and he sho fixed up fine fer 'em. Dat's de way ma and pa got married. I got married twice on Dr. Wright's place. He fixed up fer de 'casion like Marse did. Had twelve waiters both times. We had supper in de kitchen and den had dancing and music. Dem dat got married back den sho did have it in high fashion. Man would have a good striped suit, and de woman have silk and satin clothes. Dey was married by a white preacher same as de white folks. A dinner was fixed in deir honor, too. Cose, as I say, Marse Sam's slaves was treated better dan most any I ever know'd of, and all of dem loved him, too.

State: South Carolina

Interviewee: Fleming, George

"Dar was a burying ground jes' fer de slaves and de funeral was sort of like dat of de white folks. Niggers was baptized jes' like de white people, too, and by de same preacher. I saw thirty niggers baptized at one time in de river. Dat's whar everybody was baptized, den. Now dey has a basin in de church, wid glass all 'round de top, but I 'spects it do 'bout as much good.

State: South Carolina

Interviewee: Fleming, George

"During de war, food got kind of scace but didn't nobody suffer none on our place. Lawd yes, we carried de farming right on while de war was gwine on. Marse Sam's boys went to de war, but dey come back all right. Dey sho had a home-coming time fer 'em when dey got back. I heard 'bout de Yankees coming through and 'stroying things, but I never seed none. Our place stood jes' like it was all enduring de war. I didn't see no Ku Klux, neither. When freedom come, Marse called all de slaves up to de big house and say, 'I wants to know what you all is gwine to do now, fer you is free to go if you wants to.' Everybody spoke alike, 'We wants to stay wid Marse.' Everyone of de slaves stayed right on wid Marse Sam till dey could git a place to go to. Lots of 'em stayed till dey died. He divided de land up in patches and give each one a third of what was made.

State: South Carolina

Interviewee: Fleming, George

"Soon atter de war dar was a lot of trouble 'bout voting fer de governor. Some folks (like old Joe Crews) tried to put in de niggers' heads to vote fer de Republicans, but I know'd better. I voted fer Hampton like Marse did. Fact is, I voted twice fer him. (Joe Crews, and other scalawags like him,) Some scalawags had done made all de money dey could off selling niggers, so dey thought dey could make some more by making 'greements wid de Republicans. My daddy, Bill, was bullheaded. He done got dem ideas in his head and he said he gwine to vote fer de Republicans in spite of hell.

State: South Carolina

Interviewee: Fleming, George

"De Democrats done got scared 'cause so many niggers gwine to vote fer de other side, so dey formed a s'ciety called de Red Shirts. Dat was jes' to scare de niggers frum coming to de polls. I was young, but I jined right up wid dem and wore a red shirt, too.

State: South Carolina

Interviewee: Fleming, George

"Dey had a reg'lar battle in Laurens when de voting started. All de Republican niggers had deir guns stored in Tin Pot Alley, fer

State: South Carolina

Interviewee: Fleming, George

Joe Crews told 'em dey couldn't bring 'em to de polls. He thought de Yankees would protect de niggers, but fact is, de Yankees done been paid off by de Democrats and left town. Us Democrats broke in de storehouse in Tin Pot Alley and got every one of dem guns. De niggers names was on de stock of de guns. We sho had a hot time when dem niggers come up dar trying to vote. Dat's when my daddy got kil't. He had already been shot in de leg befo' dat, and dey called him 'cripple Bill'. Dem was de purtiest guns I ever seed. Dey click three times when de trigger was pulled back. Old Jim Crews was kil't, too, at dat time. Wash Hill was de one dat got him. He was shot at Crew's Branch. Twan't long atter dat till things begin to settle down, fer de Democrats sho did lick up dem Republicans.

State: South Carolina

Interviewee: Fleming, George

"I been married three times, first time I married Sarah Peter-son. I 'clar to goodness I sho can't 'member dat second one. Let me see, let me see --- Lonie, Lonie, oh yes, Lonie Golding. Us married in Laurens County on Dr. Wright's place whar I married de first one. She didn't live long and we didn't have no chilluns. My last wife name Elizabeth McKantz, she frum Abbeville. She cooked fer Mr. Jones. Her daddy was a white man, and she look jes' like a Indian. I jes' had one chile by de first wife, but he dead. His name was Richard. I got two chilluns by de last wife dat be living. Dat's Mattie, de oldest, and Hugh, de third one. I doesn't know whar neither one lives now. My other two dat's dead was Anna and George Anna. Yes, dey both named Anna, but de first one dead befo' de other one was born.

State: South Carolina

Interviewee: Fleming, George

"Some folks didn't like slavery, but I sho did. Mercy Lawd, we had a good time, den; heap better dan now. I been a long time gitting dis pension, and it ain't much when you gits it. Back in slavery times we didn't have no worries 'bout rent or something to eat. We had a job long as we lived, dat is if freedom hadn't come."

State: South Carolina

Interviewee: Fleming, George

(Floyd, Fletcher, Spartanburg, S. C., F.S. DuPre, Interviewer, Spartanburg, S. C., Edited by: Elmer Turnage)

Name:	**George Fleming** **[Gl?? Fleming]**
Home in 1900:	Hunter, Laurens, South Carolina
Age:	40
Birth Date:	Mar 1860
Birthplace:	South Carolina
Race:	Black
Gender:	Male
Relationship to Head of House:	Head
Father's Birthplace:	South Carolina
Mother's Birthplace:	South Carolina
Spouse's name:	Bettie Fleming
Marriage Year:	1880
Marital Status:	Married
Years Married:	20
Occupation:	View on Image
Neighbors:	View others on page

Household Members:

Name	Age
George Fleming	40
Bettie Fleming	40
Frank Fleming	16
Anna Fleming	4
Mattie Fleming	2
Claud Fleming	1

Source Citation: Year: *1900*; Census Place: *Hunter, Laurens, South Carolina*; Roll *T623_1533*; Page: *18B*; Enumeration District: *52*.

Source Information:

Ancestry.com. *1900 United States Federal Census* [database on-line]. Provo, UT, USA: Ancestry.com Operations Inc, 2004.

Original data: United States of America, Bureau of the Census. *Twelfth Census of the United States, 1900*. Washington, D.C.: National Archives and Records Administration, 1900. T623, 1854 rolls.

Description:
This database is an index to individuals enumerated in the 1900 United States Federal Census, the Twelfth Census of the United States. Census takers recorded many details including each person's name, address, relationship to the head of household, color or race, sex, month and year of birth, age at last birthday, marital status, number of years married, the total number of children born of the mother, the number of those children living, birthplace, birthplace of father and mother, if the individual was foreign born, the year of immigration and the number of years in the United States, the citizenship status of foreign-born individuals over age twenty-one, occupation, and more. Additionally, the names of those listed on the population schedule are linked to actual images of the 1900 Federal Census. Learn more...

TWELFTH CENSUS OF THE UNITED STATES.

B

SCHEDULE No. 1.—POPULATION.

State: South Carolina
County: Laurens

Township or other division of county: Hunter Township (part of)

Name of incorporated city, town, or village, within the above-named division: ×

Supervisor's District No. 4
Enumeration District No. 52
Sheet No. 18

Name of Institution: ×

Ward of city: ×

Enumerated by me on the 19th day of June, 1900, James H. Williams, Enumerator.

5669

Dwelling / Family No.	Name	Relation	Color	Sex	Month of birth	Year of birth	Age	Marital	Years married	Mother of how many children	Children living	Birthplace of person	Birthplace of father	Birthplace of mother	Occupation	Months not employed	Can read	Can write	Speaks English	Owned or rented	Free or mortgaged	Farm or house	Farm schedule No.
	— John	Son	W	M	Apr	1887	13	S				South Carolina	South Carolina	South Carolina			yes	yes	yes				
	— Lorie	Daughter	W	F	Nov	1888	11	S				South Carolina	South Carolina	South Carolina			yes	yes	yes				
	Finley Paul	Boarder	W	M	Aug	1877	22	S				South Carolina	South Carolina	South Carolina	Farm Laborer	0	yes	yes	yes				
[illegible]	Irby Samuel	Head	B	M	Sept	1844	55	M	30			South Carolina	South Carolina	South Carolina	Farm	0	No	No	yes	R		F	261
	— Sarah	Wife	B	F	May	1850	50	M	30	11	11	South Carolina	South Carolina	South Carolina	Farm Laborer	2	No	No	yes				[illegible]
	— Mary	Daughter	B	F	Aug	1874	25	S				South Carolina	South Carolina	South Carolina	Farm Laborer	1	yes	No	yes				
	— Fany	Daughter	B	F	Jan	1877	23	S				South Carolina	South Carolina	South Carolina	Farm Laborer	1	yes	No	yes				
	— Mariah	Daughter	B	F	Mar	1879	21	S				South Carolina	South Carolina	South Carolina	Farm Laborer	1	yes	No	yes				
	— Lize	Daughter	B	F	Feb	1881	19	S				South Carolina	South Carolina	South Carolina	Farm Laborer	1	yes	No	yes				
	— Susan	Daughter	B	F	Apr	1883	17	S				South Carolina	South Carolina	South Carolina	Farm Laborer	1	No	No	yes				
	— Sarah	Daughter	B	F	Nov	1884	15	S				South Carolina	South Carolina	South Carolina	Farm Laborer	2	No	No	yes				
	— William	Son	B	M	Sept	1886	13	S				South Carolina	South Carolina	South Carolina	Farm Laborer	2	No	No	yes				
	— Nina	Daughter	B	F	Jan	1889	11	S				South Carolina	South Carolina	South Carolina	Farm Laborer	2	No	No	yes				
	— Minnie	Daughter	B	F	May	1891	9	S				South Carolina	South Carolina	South Carolina									
	Irby Rachiel	Sister	B	F	May	1860	40	S				South Carolina	South Carolina	South Carolina	Farm Laborer	1	[illegible]	[illegible]	[illegible]				
[illegible]	Hutchinson Thomas	Boarder	B	M	Jan	1881	11	S				South Carolina	South Carolina	South Carolina	Farm Laborer	0	yes	No	yes				
[illegible]	Blakely Lawrence	Head	B	M	Nov	1846	53	M	18			South Carolina	South Carolina	South Carolina	Farm	0	No	No	yes	R		F	26[illegible]
	— Lucy	Wife	B	F	Sept	1860	[illegible]	M	18	5	5	South Carolina	South Carolina	South Carolina	Farm Laborer	1	No	No	yes				
	— Nannie	Daughter	B	F	Oct	1886	13	S				South Carolina	South Carolina	South Carolina	Farm Laborer	1	yes	No	yes				
	— Leonora	Daughter	B	F	May	1889	11	S				South Carolina	South Carolina	South Carolina	Farm Laborer	1	yes	No	yes				
	— Sallie	Daughter	B	F	Aug	1890	9	S				South Carolina	South Carolina	South Carolina									
[illegible]	— Fannie	Daughter	B	F	Jan	1893	7	S				South Carolina	South Carolina	South Carolina									
[illegible]	Simpson Lizzie	Head	B	F	May	1855	45	S		1	1	South Carolina	South Carolina	South Carolina	Day Laborer	2	yes	yes	yes			H	
	— May	Daughter	B	F	Sept	1888	12	S				South Carolina	South Carolina	South Carolina	Day Laborer	2	yes	[illegible]	[illegible]				
[illegible]	Sandifer Rufus	Head	B	M	Jan	1861	39	M	22			South Carolina	South Carolina	South Carolina	Farm	0	[illegible]	No	yes	R		F	26[illegible]
	— Mariah	Wife	B	F	Mar	1862	38	M	22	9	7	South Carolina	South Carolina	South Carolina	Farm Laborer	2	No	No	yes				
	— Lucy	Daughter	B	F	Feb	1882	18	S				South Carolina	South Carolina	South Carolina	Farm Laborer	1	yes	yes	yes				
	— Bob	Son	B	M	Apr	1883	17	S				South Carolina	South Carolina	South Carolina	Farm Laborer	0	yes	No	yes				
	— Mary	Daughter	B	F	Sept	1886	13	S				South Carolina	South Carolina	South Carolina	Farm Laborer	1	No	No	yes				
	— Martha	Daughter	B	F	Apr	1890	10	S				South Carolina	South Carolina	South Carolina									
	— Linda	Daughter	B	F	Aug	1891	8	S				South Carolina	South Carolina	South Carolina									
	— John	Son	B	M	May	1894	6	S				South Carolina	South Carolina	South Carolina									
	— George	Son	B	M	Dec	1897	2	S				South Carolina	South Carolina	South Carolina									
[illegible]	James Milton	Head	B	M	May	1858	42	M	21			South Carolina	South Carolina	South Carolina	Farm	0	yes	No	yes	R		F	264
	— Hattie	Wife	B	F	Jan	1860	40	M	20	3	3	South Carolina	South Carolina	South Carolina	Farm Laborer	1	No	No	yes				
	— Hattie	Daughter	B	F	Apr	1884	16	S				South Carolina	South Carolina	South Carolina	Farm Laborer	1	yes	No	yes				
	— Marda	Daughter	B	F	May	1896	4	S				South Carolina	South Carolina	South Carolina									
[illegible]	— Annie	Daughter	B	F	Feb	1898	2	S				South Carolina	South Carolina	South Carolina									
[illegible]	Galen [illegible]	Head	W	M	Jan	1857	43	M	12			South Carolina	South Carolina	South Carolina	Farm	0	yes	yes	yes	O	F	F	26[illegible]
	— Olive	Wife	W	F	[illegible]	1866	[illegible]	M	[illegible]	3		South Carolina	South Carolina	South Carolina			yes	yes	yes				
[illegible]	Fleming Light	Boarder	B	M	Oct	1881	18	S				South Carolina	South Carolina	South Carolina	Farm Laborer	0	yes	No	[illegible]				
[illegible]	Fleming George	Head	B	M	Mar	1860	40	M	20			South Carolina	[illegible]	South Carolina	Farm	0	No	No		R		F	26[illegible]
	— Bettie	Wife	B	F	Feb	1860	40	M	20	8	8	South Carolina	South Carolina	South Carolina	Farm Laborer	1	No	No	yes				
	— Frank	Son	B	M	Nov	1883	16	S				South Carolina	South Carolina	South Carolina	Farm Laborer	0	yes	No	yes				
	— Snow	Daughter	B	F	Sep	1888	11	S				South Carolina	South Carolina	South Carolina									
	— Mattie	Daughter	B	F	May	1898	2	S				South Carolina	South Carolina	South Carolina									
[illegible]	— Claud	Son	B	M	Apr	1899	1	S				South Carolina	South Carolina	South Carolina									
[illegible]	Anderson Willie	Head	B	M	Jan	1863	37	M	[illegible]			South Carolina	South Carolina	South Carolina	Farm	0	No	No	yes	R		F	26[illegible]
	— Allie	Wife	B	F	Mar	1873	27	M	[illegible]	2	2	South Carolina	South Carolina	South Carolina	Farm Laborer	1	yes	No	yes				
	— Noah	Son	B	M	May	1891	9	S				South Carolina	South Carolina	South Carolina									

All *Slave Narratives* results for *Mckantz*

Searching for...

Name: McKantz

More: South Carolina, USA

Edit Search

or Start a new search

Narrow by Collection

All Categories

Stories, Memories & Histories

Family Histories, Journals & Biographies

Slave Narratives

ABOUT *SLAVE NARRATIVES*

Impressive collection of interviews of former slaves in the United States

Learn more about this database

Hot Keys

New search

Refine search

Preview current record

Highlight next record

Highlight previous record

Matches **51,836-51,845** of **80,836**

State: South Carolina

Interviewee: Campbell, Thomas

"Good mornin' Marster Wood! Marster Donan McCants and Marster Wardlaw McCants both been tellin' me dat how you wants to see me but I's been so poorly and down at de heels, in my way of feelin', dat I just ain't of a mind or disposition to walk up dere to de town clock, where they say you want me to come. Take dis bench seat under de honey suckle vine. It shade you from de sun. It sho' is hot! I's surprise dat you take de walk down here to see a onery old man lak me.

State: South Carolina

Interviewee: Campbell, Thomas

"Yes sir, I was born, 'cordin' to de writin' in de Book, de 15th day of March, 1855, in de Horeb section of Fairfield District, a slave of old Marster John Kennedy. How it was, I don't know. Things is a little mixed in my mind. Fust thing I 'members, and dreams 'bout sometimes yet, is bein' in Charleston, standin' on de battery, seein' a big ocean of water, wid ships and their white sails all 'bout, de waves leapin' and gleamin' 'bout de flanks of de ships in de bright sunshine, thousands of white birds flyin' round and sometimes lighting on de water. My mammy, her name Chanie, was a holdin' my hand and her other hand was on de handle of a baby carriage and in dat carriage was one of de Logan chillun. Whether us b'long to de Logans or whether us was just hired out to them I's unable to 'member dat. De slaves called him Marster Tom. Us come back to Fairfield in my fust childhood, to de Kennedy's.

State: South Carolina

Interviewee: Campbell, Thomas

"Marster John Kennedy raise more niggers than he have use for; sometime he sell them, sometime he hire them out. Him sell mammy and me to Marster James B. McCants and I been in de McCants family ever since, bless God! "Marse James was a great lawyer in his day. I was his house boy and office boy. When I get older I take on, besides de blackin' of his boots and shoes and sweepin' out de office, de position of carriage driver and sweepin' out de church. Marster James was very 'ligious. Who my pa was? Dat has never been revealed to me. Thank God! I never had one, if they was lak I see nigger chillun have today. My white folks was all de parents I had and me wid a skin as black as ink. My belly was always full of what they had and I never suffer for clothes on my back or shoes on my feets.

State: South Carolina

Interviewee: Campbell, Thomas

"Does I 'members de Yankees? Yes sir, I 'member when they come. It was cold weather, February, now dat I think of it. Oh, de sights of them days. They camp all 'round up at Mt. Zion College and stable their hosses in one of de rooms. They gallop here and yonder and burn de 'Piscopal Church on Sunday mornin'. A holy war they called it, but they and Wheeler's men was a holy terror to dis part of de world, as naked and hungry as they left it. I marry Savannah Parnell and of all our chillun, dere is just one left, a daughter, Izetta. Her in Tampa, Florida.

State: South Carolina

Interviewee: Campbell, Thomas

"Does I 'members anything 'bout de Ku Klux? No sir, nothin'. I was always wid de white folks side of politics. They wasn't concerned 'bout me. Marster James have no patience for dat kind of business anyhow. Him was a lawyer and believed in lettin' de law rule in de daylight and would have nothin' to do wid work dat have to have de cover of night and darkness.

State: South Carolina

Interviewee: Campbell, Thomas

"Does I 'member 'bout de red shirts? Sure I does. De marster never wore one. Him get me a red shirt and I wore it in Hampton days. What I recollect 'bout them times? If you got time to listen. I 'spect I can make anybody laugh 'bout what happen right in dis town in red shirt days. You say you glad to listen? Well, here goes. One time in '76. de democrats have a big meetin' in de court house in April. Much talk last all day. What they say or do up dere nobody know. Paper come out next week callin' de radicals to meet in de court house fust Monday in May. Marster Glenn McCants, a lawyer, was one of old marster's sons. He tell me all 'bout it.

State: South Carolina

Interviewee: Campbell, Thomas

"De day of de radical republican meetin' in de court house, Marster Ed Ailen had a drug store, so him and Marster Ozmond Buchanan fix up four quart bottles of de fimest kind of liquor, wid croton-oil in every bottle. Just befo' de meetin' was called to order, Marster Ed pass out dat liquor to de ring leader, tellin' him to take it in de court house and when they want to 'suade a nigger their way, take him in de side jury rooms and 'suade him wid a drink of fine liquor. When de meetin' got under way, de chairman 'pointed a doorkeeper to let nobody in and nobody out 'til de meetin' was over widout de chairman say so.

State: South Carolina

Interviewee: Campbell, Thomas

"They say things went along smooth for a while but directly dat croton-oil make a demand for 'tention. Dere was a wild rush for de door. De doorkeeper say 'Stand back, you have to 'dress de chairman to git permission to git out'. Chairman rap his gavel and say; 'What's de matter over dere? Take your seats! Parliment law 'quire you to 'dress de chair to git permission to leave de hall'. One old nigger, Andy Stewart, a ring leader shouted: 'To hell wid Parliment law, I's got to git out of here.' Still de doorkeeper stood firm and faithful, as de boy on de burnin' deck, as Marster Glenn lak to tell it. One bright mulatto nigger, Jim Mobley, got out de tangle by movin' to take a recess for ten minutes, but befo' de motion could be carried out de croton oil had done its work. Half de convention have to put on clean clothes and de court house steps have to be cleaned befo' they could walk up them aga You ask any old citizen 'bout it. Him will 'member it. Ask old Doctor Buchanan. His brother, de judge, was de one dat help Marster Ed Aiken to fix de croton-oil and whiskey.

State: South Carolina

Interviewee: Campbell, Thomas

"Well, dat seem to make you laugh and well it might, 'cause dat day been now long ago. Sixty-one years you say? How time gits along. Well, sixty-one years ago everybody laugh all day in Winnsboro, but Marster Ed never crack a smile, when them niggers run to his drug store and ask him for somethin' to ease their belly ache."

State: South Carolina

Interviewee: Cannon, Sylvia

Cannon, Sylvia

[redacted]

Name:	**James B Mccants**
Birth Year:	abt 1818
Age in 1870:	52
Birthplace:	South Carolina
Home in 1870:	Township 4, Fairfield, South Carolina
Race:	White
Gender:	Male
Value of real estate:	View image
Post Office:	Winnsboro

Household Members:

Name	Age
James B Mccants	52
Laura Mccants	43
George B Mccants	24
Sarah Mccants	22
Glenn Mccants	20
Laura Mccants	8

Source Citation: Year: *1870*; Census Place: *Township 4, Fairfield, South Carolina*; Roll *M593_1496*; Page: *93B*; Image: *190*; Family History Library Film: *552995*.

Source Information:

Ancestry.com. *1870 United States Federal Census* [database on-line]. Provo, UT, USA: Ancestry.com Operations, Inc., 2009. Images reproduced by FamilySearch.

Original data:

- 1870 U.S. census, population schedules. NARA microfilm publication M593, 1,761 rolls. Washington, D.C.: National Archives and Records Administration, n.d.
- Minnesota census schedules for 1870. NARA microfilm publication T132, 13 rolls. Washington, D.C.: National Archives and Records Administration, n.d.

Description:
This database is an index to individuals enumerated in the 1870 United States Federal Census, the Ninth Census of the United States. Census takers recorded many details including each person's name, age at last birthday, sex, color; birthplace, occupation, and more. No relationships were shown between members of a household. Additionally, the names of those listed on the population schedule are linked to actual images of the 1870 Federal Census. Learn more...

Page No. 96 } Inquiries numbered 7, 16, and 17 are not to be asked in respect to infants. Inquiries numbered 11, 12, 15, 16, 17, 18, and 20 are to be answered (if at all) merely by an affirmative mark, as /.

SCHEDULE 1.—Inhabitants in Township No 4, **in the County of** Fairfield, **State of** South Carolina, **enumerated by me on the** 11th **day of** June, **1870.**

Post Office: Winnsboro — B. J. Dawson, Ass't Marshal.

Line	1 Dwelling	2 Family	3 Name	4 Age	5 Sex	6 Color	7 Profession, Occupation, or Trade	8 Value of Real Estate	9 Value of Personal Estate	10 Place of Birth	11	12	13	14	15	16	17	18	19	20
1			O'Ulin Catherine	2	F	W				South Carolina										
2			Grace	9/12	F	W				South Carolina			Sept							
3			Stewart Martha	34	F	W	At home			South Carolina										
4			Ferguson James	17	M	W	Farm laborer			South Carolina										
5	252	252	Jenkins Henry	34	M	B	Farmer		175	South Carolina						1	1		1	✓
6			Margaret	30	F	B	Farm laborer			South Carolina						1	1			
7	253	253	Le Kee Daniel	35	M	B	Farmer		155	South Carolina						1	1		1	✓
8			Rosa	30	F	B	Farm laborer			South Carolina						1	1			
9			Samuel	15	M	B	Farm laborer			South Carolina						1	1			
10	254	254	Edwards Caroline	35	M	B	Farmer		165	South Carolina						1	1		1	✓
11			Amelia	55	F	B	Farm laborer			South Carolina						1	1			
12			Eugenia	12	F	B	Farm laborer			South Carolina						1	1			
13			William	10	M	B				South Carolina										
14			Ezekiel	9/12	M	B				South Carolina			Dec							
15			Eliza	25	F	B	Domestic servant			South Carolina						1	1			
16	255	255	McCants James	52	M	W	Attorney at law	15000	35000	South Carolina									1	✓
17			Laura	43	F	W	Keeping house			South Carolina										
18			George W	24	M	W	Farmer			South Carolina									1	✓
19			Sarah	22	F	W	At home			South Carolina										
20			Glenn	20	M	W	Teaching school			South Carolina										
21			Laura	8	F	W				South Carolina					1					
22	256	256	McDuffie Lavinia	50	F	B	Domestic servant			South Carolina						1	1			
23			Ashley	14	F	B	Domestic servant			South Carolina							1			
24			Aiken Howard	15	M	B	Domestic servant			South Carolina							1			
25			Coulter Celia	60	F	B	Domestic servant			South Carolina						1	1			
26			Scott Milly	13	F	B	Domestic servant			South Carolina						1	1			
27	257	257	Woodward Gustavus A	31	M	W	Teaching school		275	South Carolina									1	✓
28			Charlotte E	25	F	W	Keeping house			South Carolina										
29			John	6	M	W				South Carolina										
30			Edward O	4	M	W				South Carolina										
31			Baby	5/12	M	W				South Carolina			Dec							
32	258	258	Jackson Milly	55	F	M	Domestic servant			South Carolina						1	1			
33			Hiram	16	M	B	Day laborer			South Carolina							1			
34			Jeremiah	15	M	B	Day laborer			South Carolina							1			
35			Lyles Quimby	55	F	B	Domestic servant			South Carolina						1	1			
36			Adeline	16	F	M	Domestic servant			South Carolina							1			
37	259	259	Rion James H	42	M	W	Attorney at law	7500	5000	South Carolina									1	✓
38			Mary E	41	F	W	Keeping house			South Carolina										
39			Preston	18	M	W	Attending school			South Carolina					1					
40			Margaret	16	F	W	At home			South Carolina					1					

No. of dwellings, 8; No. of white females, 9; No. of males, foreign born, —; 22,510; 43,530; No. of insane, —

No. of families, 8; No. of colored males, 9; No. of females, foreign born, —; 3; 3; 14; 7

No. of white males, 10; No. of colored females, 12; No. of blind, —

21

WORLD WAR II
REGISTRATION CARDS

REGISTRATION CARD—(Men born on or after April 28, 1877 and on or before February 16, 1897)

SERIAL NUMBER	1. NAME (Print)			ORDER NUMBER
U 1231	Henry (First)	(Middle)	Patterson (Last)	

2. PLACE OF RESIDENCE (Print)
1943 N. Dearin St. (Number and street) Phila. (Town, township, village, or city) (County) Pa. (State)

[THE PLACE OF RESIDENCE GIVEN ON THE LINE ABOVE WILL DETERMINE LOCAL BOARD JURISDICTION; LINE 2 OF REGISTRATION CERTIFICATE WILL BE IDENTICAL]

3. MAILING ADDRESS
Same
[Mailing address if other than place indicated on line 2. If same insert word same]

4. TELEPHONE	5. AGE IN YEARS	6. PLACE OF BIRTH
No	62	Greenwood (Town or county)
(Exchange) (Number)	DATE OF BIRTH: March 17 1880 (Mo.) (Day) (Yr.)	South Carolina (State or country)

7. NAME AND ADDRESS OF PERSON WHO WILL ALWAYS KNOW YOUR ADDRESS
James Patterson 1943 N. Dearin St. Phila, Pa.

8. EMPLOYER'S NAME AND ADDRESS
Disston Saw Works

9. PLACE OF EMPLOYMENT OR BUSINESS
Tacony (Number and street or R. F. D. number) Phila. (Town) (County) Pa. (State)

I AFFIRM THAT I HAVE VERIFIED ABOVE ANSWERS AND THAT THEY ARE TRUE.

D. S. S. Form 1 (Revised 4-1-42) (over) 16—21630-2 Henry Patterson (Registrant's signature)

REGISTRAR'S REPORT

DESCRIPTION OF REGISTRANT

RACE		HEIGHT (Approx.)	WEIGHT (Approx.)	COMPLEXION	
White	X	5-6	116	Sallow	
Negro		EYES	HAIR	Light	
Oriental		Blue	Blonde	Ruddy	
Indian		Gray	Red	Dark	X
Filipino		Hazel	Brown	Freckled	
		Brown X	Black X	Light brown	
		Black	Gray	Dark brown	
			Bald	Black	

Other obvious physical characteristics that will aid in identification

I certify that my answers are true; that the person registered has read or has had read to him his own answers; that I have witnessed his signature or mark and that all of his answers of which I have knowledge are true, except as follows:

E.W. Armstrong (Signature of registrar)

Registrar for Local Board Indiana, Pa. (Number) (City or county) (State)

Date of registration April 27 1942

LOCAL BOARD NO. 9
INDIANA COUNTY
143 Indiana Theatre Bldg.
INDIANA, PENNA.

(STAMP OF LOCAL BOARD)

(The stamp of the Local Board having jurisdiction of the registrant shall be placed in the above space)

16—21630-2

REGISTRATION CARD—(Men born on or after April 28, 1877 and on or before February 16, 1897)

SERIAL NUMBER	1. NAME (Print)			ORDER NUMBER
U 2373	Purvis (First)	(Middle)	Patterson (Last)	

2. PLACE OF RESIDENCE (Print)

4740 State (Number and street) Chicago (Town, township, village, or city) Cook (County) Ill (State)

[THE PLACE OF RESIDENCE GIVEN ON THE LINE ABOVE WILL DETERMINE LOCAL BOARD JURISDICTION; LINE 2 OF REGISTRATION CERTIFICATE WILL BE IDENTICAL]

3. MAILING ADDRESS

4740 State St. Chicago Ill C

(Mailing address if other than place indicated on line 2. If same insert word same)

4. TELEPHONE	5. AGE IN YEARS	6. PLACE OF BIRTH
	54	Greenwood (Town or county)
(Exchange) (Number)	DATE OF BIRTH: July (Mo.) 4 (Day) 1887 (Yr.)	South Carolina (State or country)

7. NAME AND ADDRESS OF PERSON WHO WILL ALWAYS KNOW YOUR ADDRESS

Trannie Patterson 4740 State St.

8. EMPLOYER'S NAME AND ADDRESS

Eddie Mathews Decorator

9. PLACE OF EMPLOYMENT OR BUSINESS

5356 Dearborn St (Number and street or R.F.D. number) Chicago Ill (Town) (County) Ill. (State)

I AFFIRM THAT I HAVE VERIFIED ABOVE ANSWERS AND THAT THEY ARE TRUE.

Purvis Patterson
(Registrant's signature)

D. S. S. Form 1
(Revised 4-1-42) (over) 16—21630-2

PICTURE ALBUM

DESCENDANTS

OF

RUBEN PATTERSON

nearly 10,000 free blacks. By the mid-18th century the slaves on rice plantations provided their **masters with the highest per capita income** in the American colonies. Many of the Africans who were brought to the South Carolina low country came from rice-producing areas of Africa. African methods of planting, hoeing, winnowing, and threshing rice were used as late as 1865.

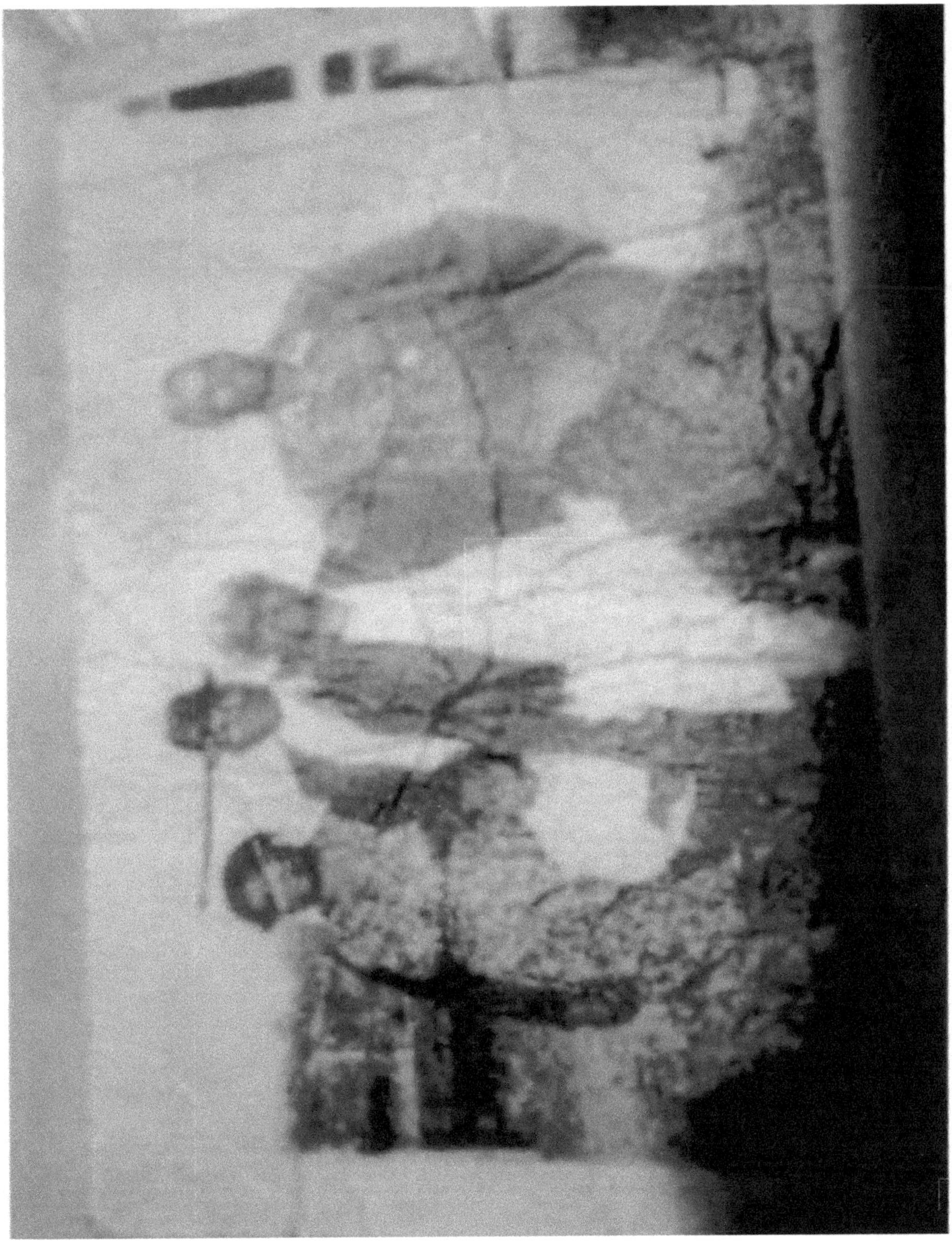

Left to Right: Rube Patterson Children: Willie Mae Patterson Hill, Robert Patterson, Mamie Lee Marshall, Humphrey Marshall (son-in-law, and Great Grandson.

Elbert Patterson

Susie Holloway Wade, Dorothy A. Chappelle, Mozanna Clinkscale

Cora
Emma
Corrine
Dorothy
Grand daughter

Carrie
Mamie Lee
Will
David
Ruben
Louis
DESCENDANTS
OF
WILLIE ANN HOLLOWAY & RUBEN PATTERSON
Robert
Willie Mae
Evelyn
Edd
Cora Emma
Kathleen

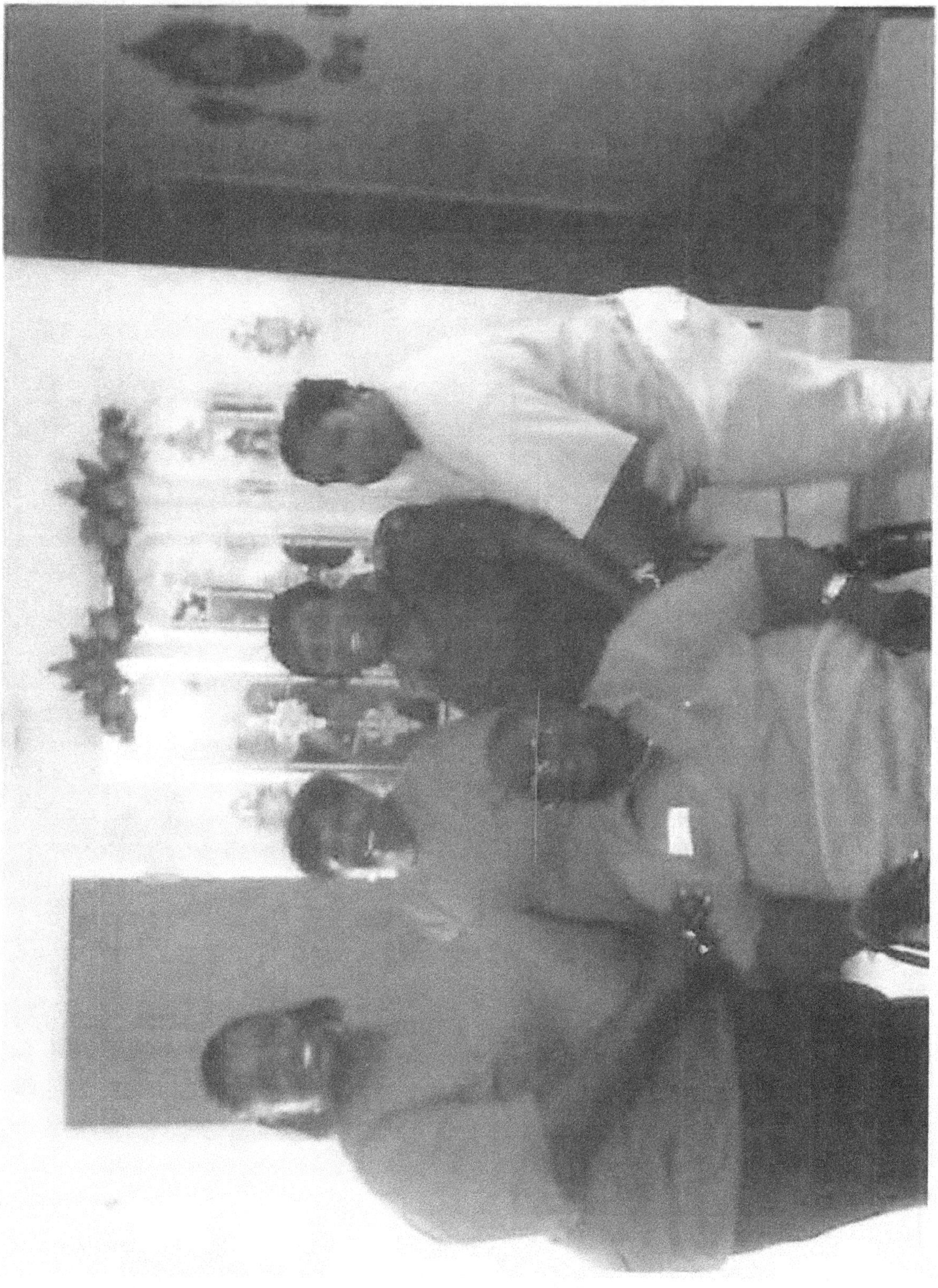

Dorothy's Children

Left to Right: Cynthia A. Chappelle, Dorothy A. Chappelle (Mom), Elaine Williams, Tonya Chappelle, Keisha Patterson

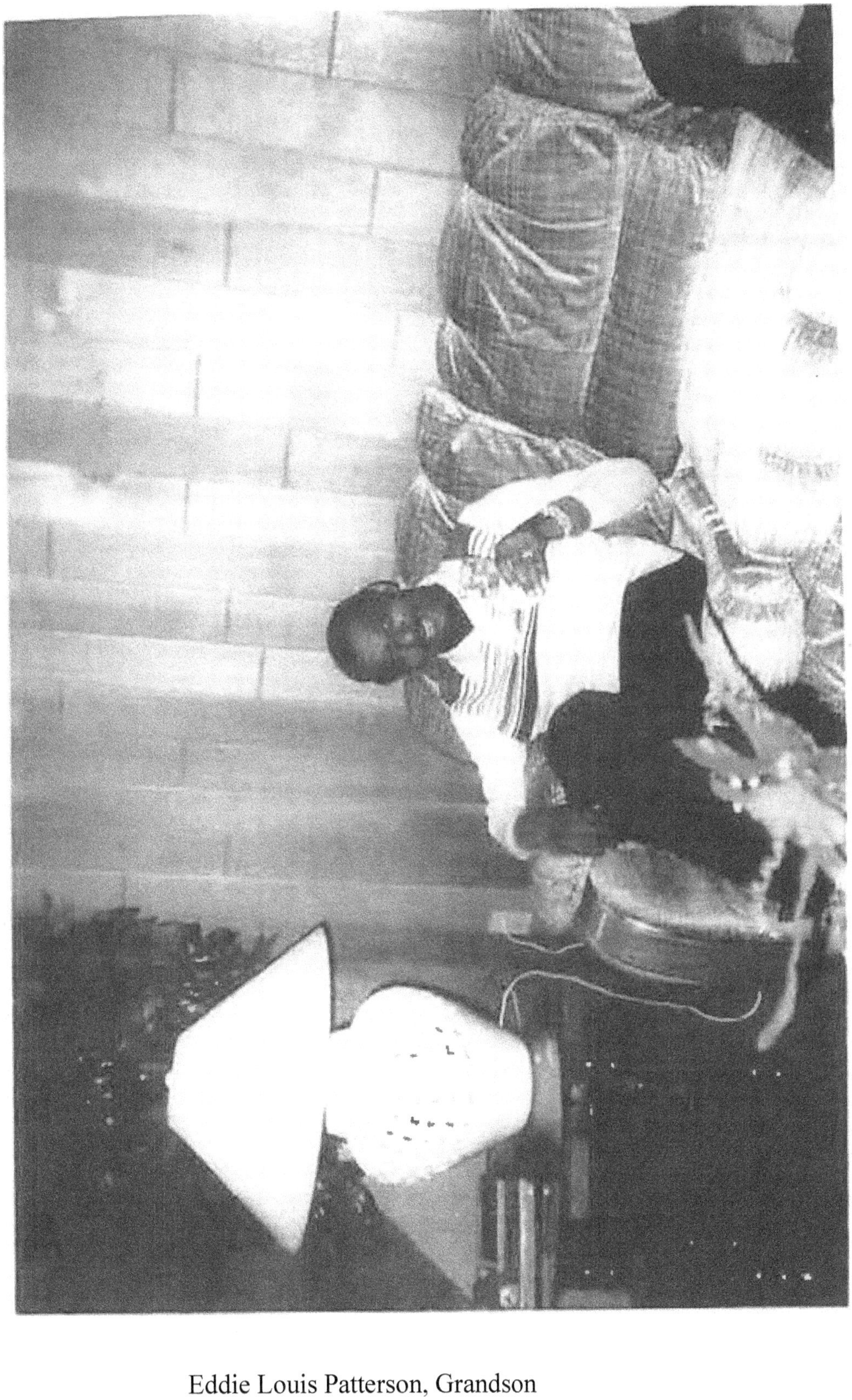

Eddie Louis Patterson, Grandson

Desendant of Ruber Patterson
Willie Mae
Daughter
Carrine
Grand daughter

Desendants of Ruben Patterson

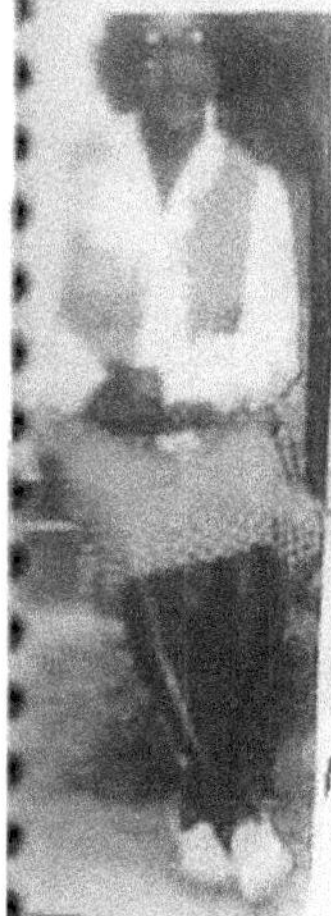

Mattie & Ruben

Theodore,

Johnny

Ricky

Michael

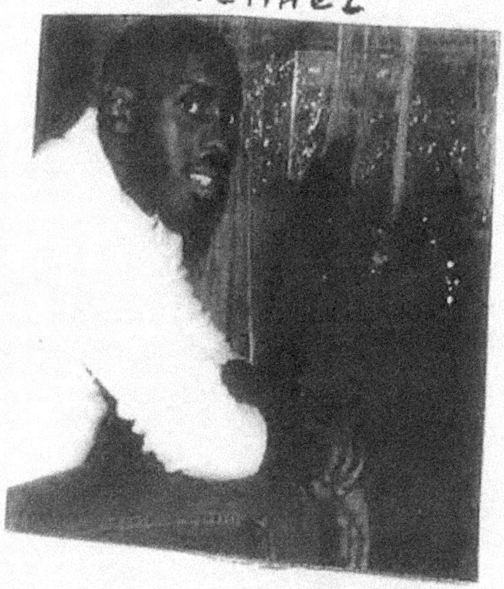

Mike 1969 5th grade

Theodore Michael Louis 1990

Descendants of Ruben Patterson

Kathleen And Thomas Brooks Family

Grandson Shawn "Bo"

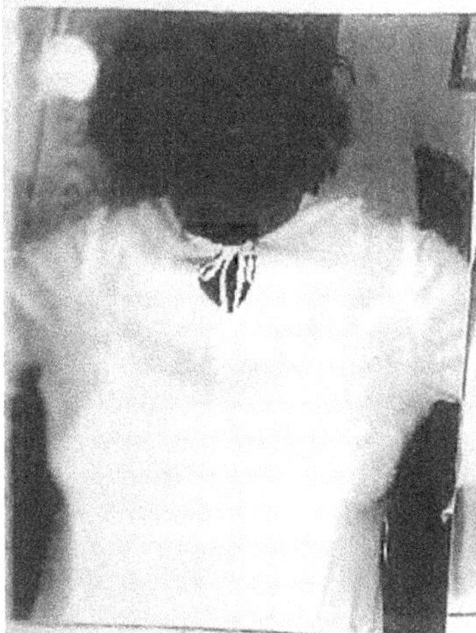

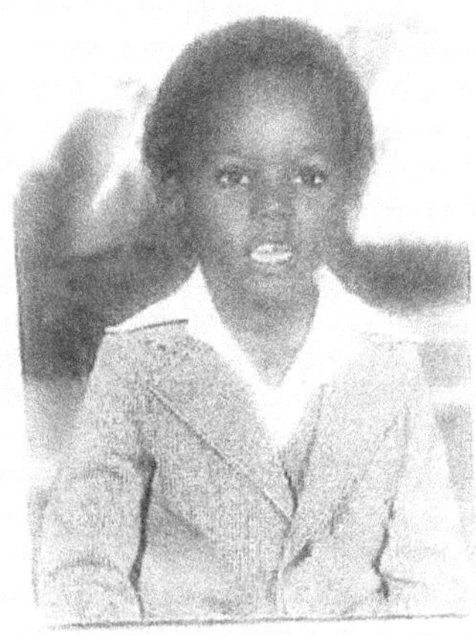

Kat

Thomas

7th Grade

Barry

Darlene

Ernestine

Melvin

Cathy, Tira, Teresa

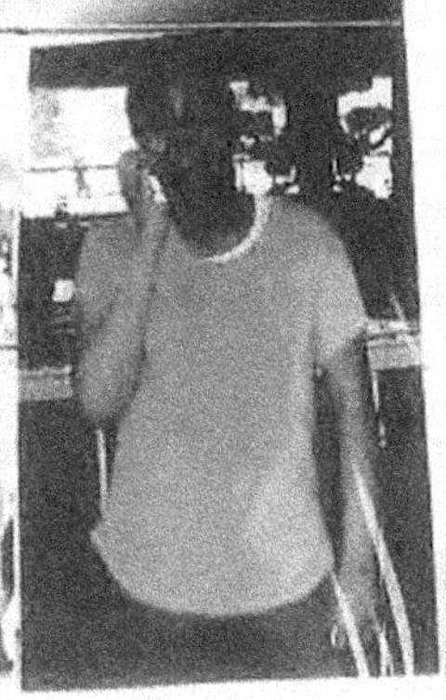

Desendants of Ruben Patterson

Desendants of Reuben Patterson

Anam Ult- Linda K Marshall . Edward Marshall - Kenneth

Desendants of Ruben Patterson

Daniel ("Bay") & Cleo Potter
Cleo
Ruby
Nathenial
Bobby
John Allen
Ann
(Granddaughter)
Jennymae
Daisey "p"

Patterson family is Wince Patterson Childrens of Ga.

Dean-Patterson Family Picture Day

Greenwood (Ninety Six), SC

& Picnic

Augusta, Georgia

May 30, 2009

Franklin Park
2501 Franklin Dr Ft. Lauderdale 33311
12:00 p.m.—5:00 p.m.
$5.00 per person 6yrs and up

Do not miss your chance to meet family members you never knew. Many of you Will see people you know but did not know you were related. We're going to have a wonderful time.

Additional Family Photo Project

We would like to put together a family photo dvd. We are asking head family members to collect a photo of each member in their family (one person per photo). Please put the name of the person in the photo, as well as, who the photo belongs to. Once photos have been collected, place all photos in one envelope. Give envelop to Danielle or Zandra 954.584.4528 either at the next meeting or bring them to the picnic. **All photos are due by June 5, 2009.**

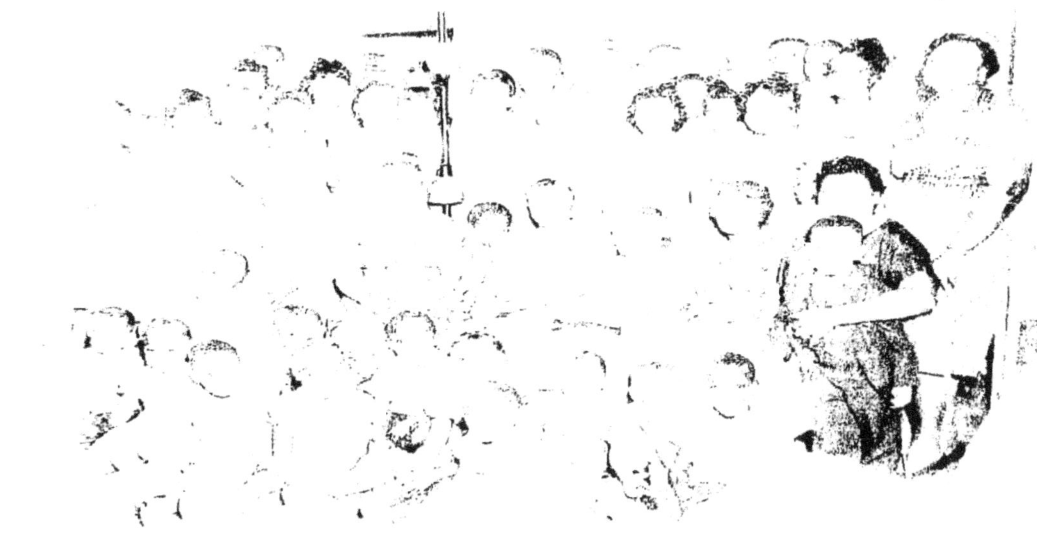

During the picnic we would like to take a family picture. We are trying to include as many family members as possible. We have not had a group photo for many generations. Please encourage ALL loved-ones to come and be a part of this special moment. Please contact Patricia Hankerson with questions 954-527-0414

PATTERSON FAMILY REUNION 2005

"We Are Family"

Friday, July 1, 2005 - Sunday, July 3, 2005

Hampton Inn University Place
8419 North Tryon Street
Charlotte, NC 28262

The Patterson Story

Many families have lost their ancestral roots over the passing years. This task truly has been difficult for us in the the Black Community because of the half-hearted recordkeeping that took place. Many Blacks were listed in the United States Census as Black Girl or Black Boy born to Mr. So and So. Before the 1860 Census, Blacks were listed (if at all) as Indians. With no dates to identify the ages or names, much of our history is lost. The Blacks were considered as only property. With limited education, no knowledge of how to read and write, there is very little of our history written on paper. Fortunately, God produced great story tellers. We have ancestors that told the story verbally of who our ancestors were. How they lived, how they made a living. How they made something out of nothing. We are proud to try and recap the Patterson Family Story through this book.

Born in the South during slavery was Pink Patterson, probably the property of Master Joe Patterson. After he was freed, he changed his name to Willis Patterson. He had four siblings (Wince, Nelson, Elbert, and Jane) that we know about; there may be more that were possibly sold into slavery. As a customary procedure during that time, Willis probably had to ask his slave owner for permission to marry Nancy McKinnsey. Pink (Willis) Patterson and Nancy McKinnsey began their life with very little. As the United States had promised each slave 40 acres and a mule, they were given nothing to begin their lives as free men. With no place to go, many of our ancestors remained on the Master's property and worked for him for peanuts (limited pay). This fact did not escape Willis and Nancy as they worked and lived on the Kinard place for some years.

With hopes and dreams for a better life for their children, Willis and Nancy trusted God to provide for their family. As God led him, their son, Ruben Patterson worked hard and tireless hours to move his family to another home in Greenwood, South Carolina. He instilled courage, pride, and Christian values in his children. The Patterson Legacy was passed down to their eleven children: Carry, Mamie Lee, Will (Buddy), David (Bay), Ruben (Sonny Boy), Louis (Duise), Robert (Pop & TL), Willie Mae, Evelyn, Edd, and Cora Emma (Sis). This is their story.

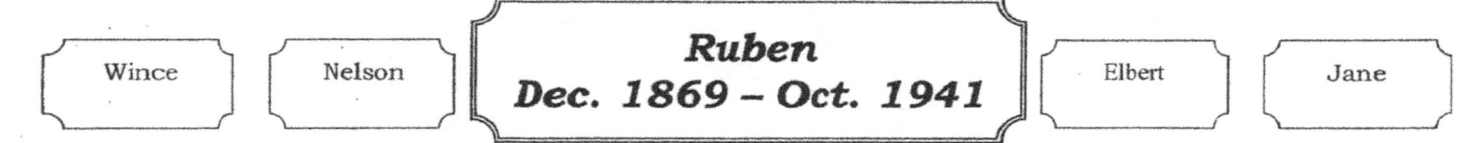

Wince | Nelson | **Ruben** | Elbert | Jane

Ruben
Dec. 1869 – Oct. 1941

Born in December, 1869 Ruben Patterson was one of five children *(Wince, Nelson, Elbert and Jane)* that were conceived by Willis Patterson and Nancy McKinnsey. At a young age he joined Old Mount Zion Baptist in Greenwood County. He married the lovely Willie Ann Holloway and produced 11 children and raised 4 grandchildren. Ruben and Willie Ann both lived and worked on Kinard Place. A hard working, quiet man that farmed plants and cotton, he was committed to his wife and family. Waking early to begin his day in the fields, Paw didn't miss a day without his Luzianne Coffee served pippin hot. Paw began sharing his Luzianne coffee with his beloved grandson Louis at an early age. Recognized in the community by the patch that was worn over his eye, no one seems to now how he lost his eye. Later in life Ruben joined Beulah Baptist in Greenwood. He was committed and adamant about him and his children attending church. They later moved into the town of Greenwood and walked to Beulah to attend services regularly. Ruben finished his Christian Journey in October, 1941 and went from labor to reward. He is buried in the cemetery of his beloved church of Beulah Baptist.

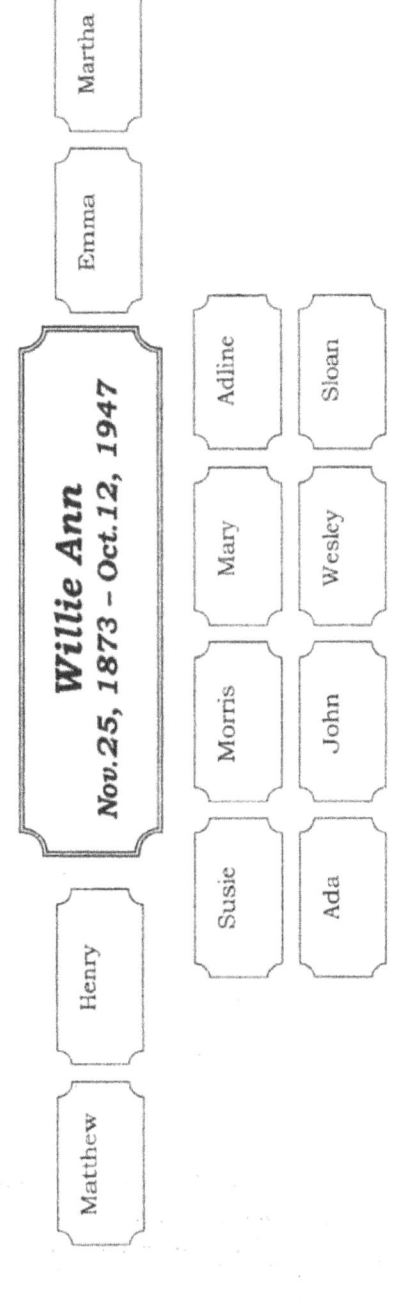

Matthew | Henry

Willie Ann
Nov.25, 1873 - Oct.12, 1947

Emma | Martha

Susie | Morris | Mary | Adline

Ada | John | Wesley | Sloan

Ten years shy of slavery Willie Ann Holloway was born to Sam Holloway and Martha Folks on November 25, 1873. One of thirteen children Willie Ann grew up on a farm in Greenwood County. Long hair and stout, Willie Ann was courted by Ruben Patterson and jumped the broom (joined in marriage) and produced 11 children. Willie Ann worked hard and strutted her stocky 6'0" frame around the kitchen in long dresses and aprons to provide hearty meals for her children which included the biggest mouth watering biscuits ever. For years Ruben and Willie Ann worked and lived on the Kinard Place. Willie Ann and Ruben moved to Martin Chin Place and she later did house keeping for the Cobb family. The family moved their membership to Beulah Baptist Church.

Willie Ann loved her children and supported them in their life. Willie Ann and Ruben had a hand in raising the grandkids that lived in their home or close by including (Kathleen and Corrine). Unsure about the Promise land of the North, Willie Ann refused to allow two of her grandchildren Dorothy Ann and Lewis to journey to that unknown place. After Ruben's death, Willie Ann, Willie Mae, Corrine, Dorothy, and Louis moved to the Cobb Place until the Lord called her home on October 12, 1947. Willie Ann and Ruben lay side by side in the Beulah Baptist Church Cemetery.

Carry

The oldest child of Ruben and Willie Ann. Carry was born in Greenwood; SC. Carry briefly attended the only Black School in Greenwood, Blakedale School. She worked hard farming along with her parents and siblings. She married a Greenwood native Mr. Pete McBride. Carry had one child, a wonderful daughter, ***Kathleen***. Carry left Greenwood for New York with hopes and dreams of earning a decent living. Once she was settled in Brooklyn, Carry quickly found employment. She worked in the Garment District and retired in that profession. She died in her early sixty's in Brooklyn and is buried in New Jersey. Her descendents are as follows:

Kathleen

Mamie Lee

Mamie Lee a tall and medium build women was the second child born unto Ruben and Willie Ann. She like her sister briefly attended Blakedale School. Mamie Lee did domestic work and farming until she married Humphrey Marshall. They lived in Hodges, SC where they raised six children. A dedicated mother and wife Mamie Lee became a homemaker and took care of her family on a full time basis. She enjoyed cooking and provided three tasty meals a day for her family which included biscuits or cornbread. Humphrey whom was determined to provide for his family was one of a few black men that owned a car and he took great pride in that fact. Their descendents are as follows:

James (Boy)

Lue Ella

Joseph

Johnny

Edd (Turkey)

Mary Lee

Will (Buddy)

Willie was the first son born to this union. Itching to see the world, Buddy left Greenwood early. He was known to travel extensively. He worked for the Railroad which aloud him to see the world and get to know a lot of people. He would often bring people home and asked his *Maw* to feed them. You could here him say"*this is my friend- we'll eat a little and move on.*" Buddy joined the Army Reserve for a brief time. When he would come home to visit, he would tell stories about his adventures. He would say "Shut up and listen, you might learn something." He was a good cook and kept a clean house. Willie returned to Greenwood and married Marie and they both worked for Dr. Bishop and resided on Dr. Bishop property. Yearning for more for his family Buddy moved his family up north to New Jersey. His first job was as a body guard for "Daddy Grace", a well prominent Preacher. He worked on this job for quite awhile and retired in Plainsfield, New Jersey. His final trip was taken in his late seventy's as he entered his eternal rest. Their descendents are as follows:

Kathleen

John Carl

David (Bay)

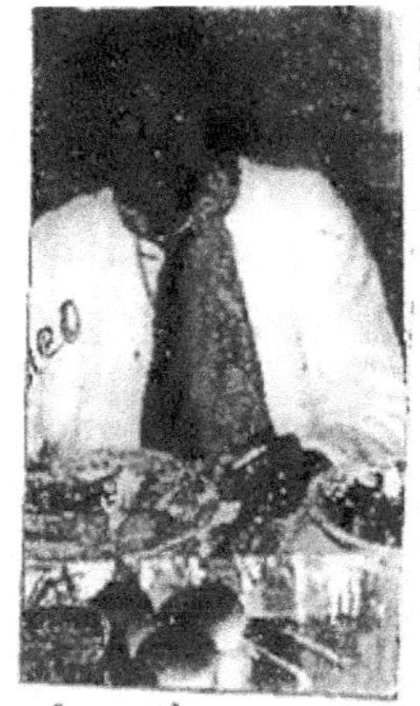

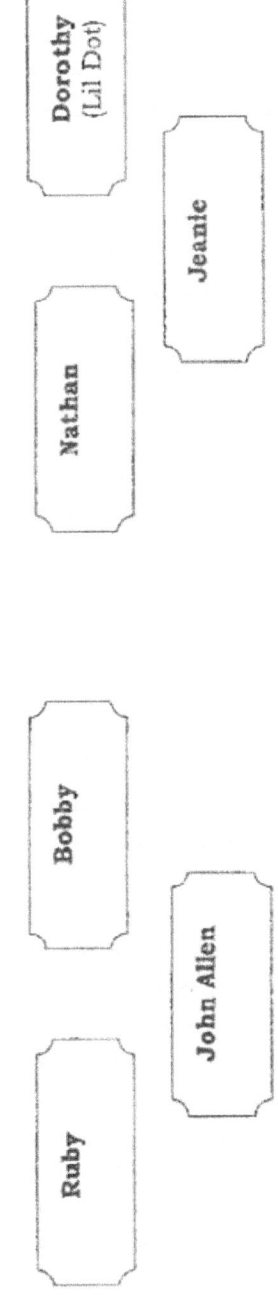

David was a very responsible man who loved his family. He did not allow his limited education to hinder him from accomplishing things in life. Remaining in Greenwood County all his life Bay used construction work and farming cotton as a tool to provide for his family. Bay was a very quiet man, but had a great sense of humor that kept every one laughing. Feeling a sense of responsibility Bay often visited his parents to check to see if they were alright and just to talk. As he talked he would often say *"You just don't know."* He married Cleo Aiken and they had six children. They raised their family in Hodges, SC. He departed his life in his early forty's in a car accident in Hodges. Their daughter Dorothy, affectionate known as "Lil Dot" died early. Their descendents are as follow:

Ruby

Bobby

Nathan

Dorothy
(Lil Dot)

John Allen

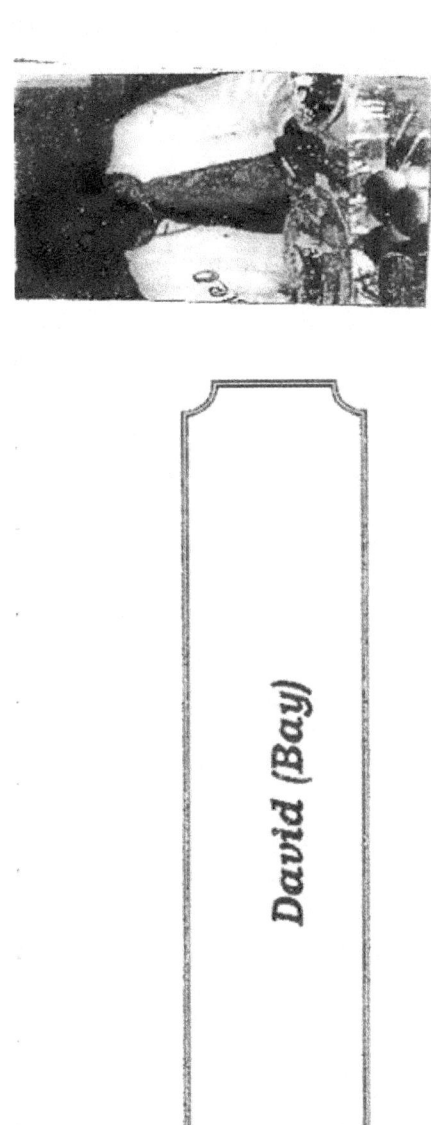

Jeanie

Ruben (Sonny Boy)

Ruben affectionately known as Sonny Boy resembled his name sake; Ruben With his tall frame, Sonny was a serious man with a great sense of humor. He worked in construction for several years and later became a plumber at the Mill in Greenwood. With a fond love for Greenwood, Sonny lived his entire life in Greenwood. He enjoyed cooking and loved being in the company of his nieces and nephews. At Christmas he would buy toys for them and play jokes on them.

He married Mattie and they had four boys. Ruben passed from labor to reward at a very early age, early forty's. Their descendents are as follows:

Theodore | **Johnny** | **Rickie** | **Michael**

Louis (Dusie)

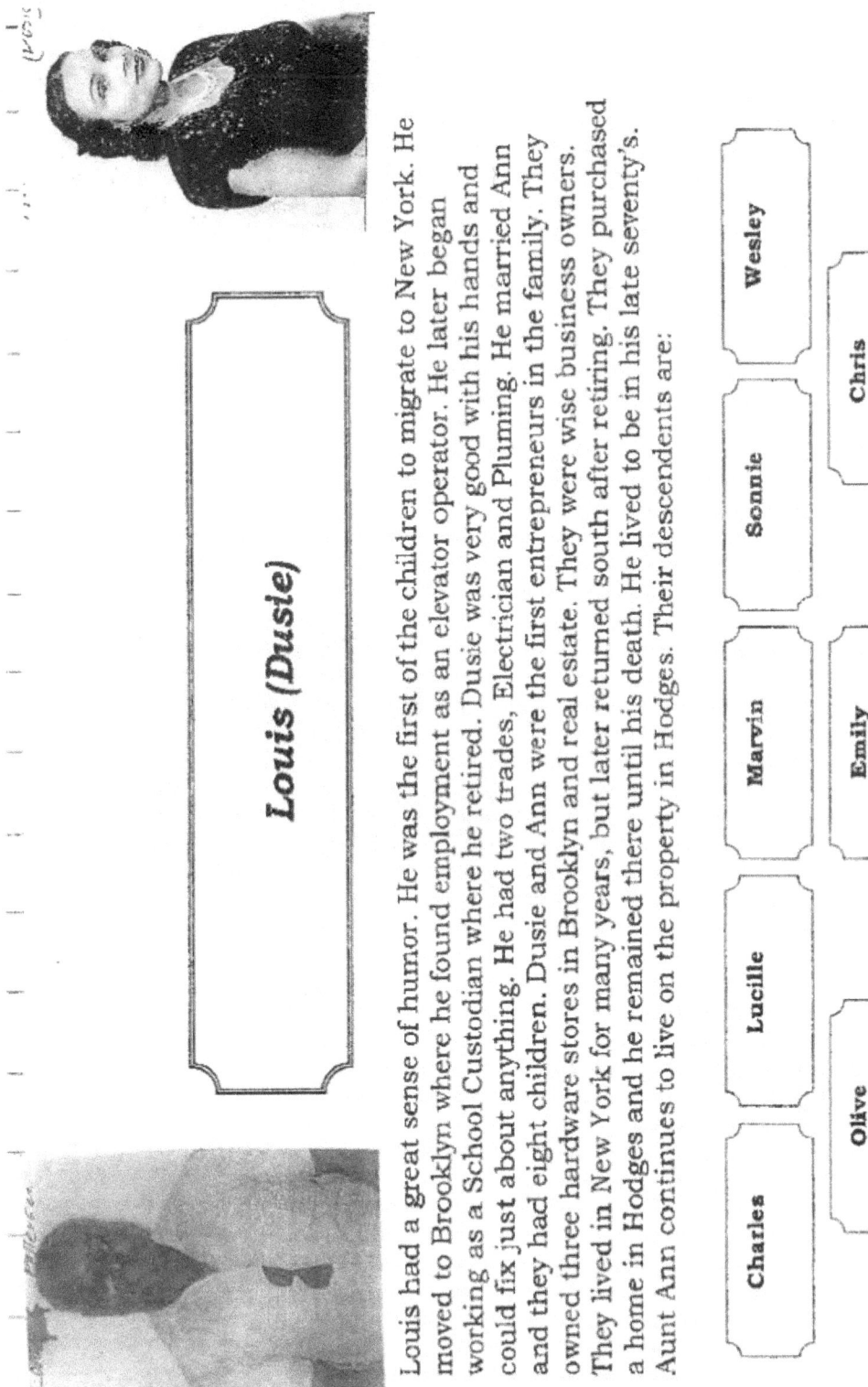

Louis had a great sense of humor. He was the first of the children to migrate to New York. He moved to Brooklyn where he found employment as an elevator operator. He later began working as a School Custodian where he retired. Dusie was very good with his hands and could fix just about anything. He had two trades, Electrician and Pluming. He married Ann and they had eight children. Dusie and Ann were the first entrepreneurs in the family. They owned three hardware stores in Brooklyn and real estate. They were wise business owners. They lived in New York for many years, but later returned south after retiring. They purchased a home in Hodges and he remained there until his death. He lived to be in his late seventy's. Aunt Ann continues to live on the property in Hodges. Their descendents are:

Charles | **Lucille** | **Marvin** | **Sonnie** | **Wesley**

Olive | **Emily** | **Chris**

Robert (Pop &TL)

Bald headed, strutting in his suspenders and smoking a cigar Robert like his sibling left the tough South for a more rewarding living in New York. He arrived in Brooklyn and began working as an elevator operator. As humor was important to the other male siblings it was also with Pop who loved to laugh. A serious man Pop was an all around guy. Even though he had a great sense of humor Pop was quiet at times. He loved getting together with his family, whether it was up north or down south. Robert married Eunice and they lived in Brooklyn until death. They did not have any children. Pop lived to be 76 years of age.

Willie Mae

With a love for family Willie Mae remained in Greenwood County all her life. As a care giver she did domestic work for most of her life. She was well respected at home, Church and Community. She loved the Lord and her Church, Beulah Baptist. She served diligently at her Church until the Lord called her home. A serious woman, who spoke her mind without fear. She was an excellent cook and baker. Her grand's would look forward to her fried chicken, and homemade cakes, pies and cookies. It was good eating unless it was made from scratch. She married Jim Hill. Willie Mae lived to be 76 years of age. Her descendants are as follow:

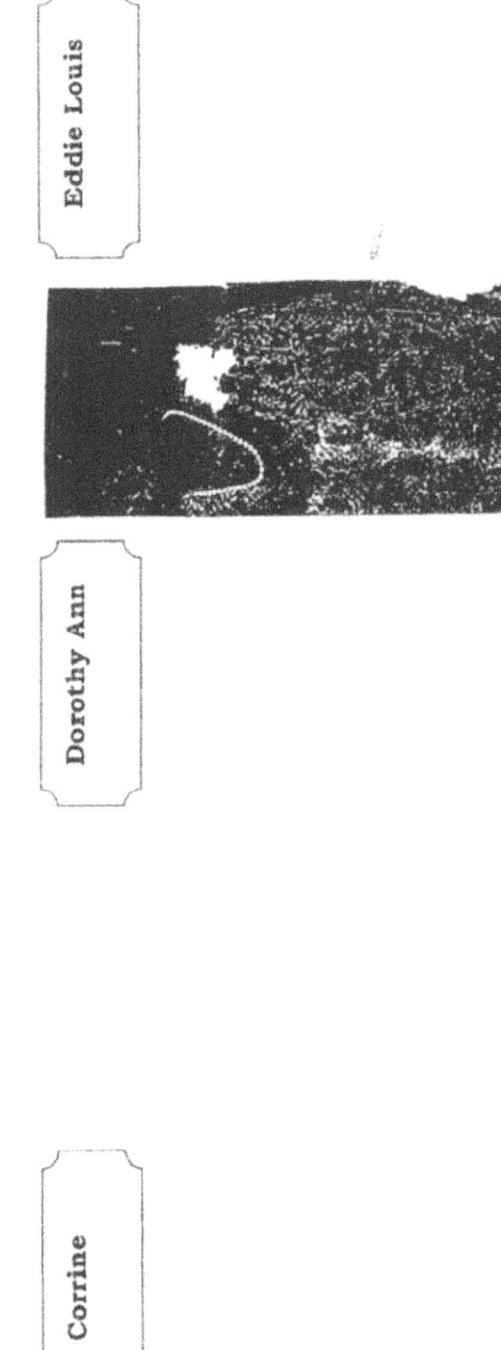

Corrine

Dorothy Ann

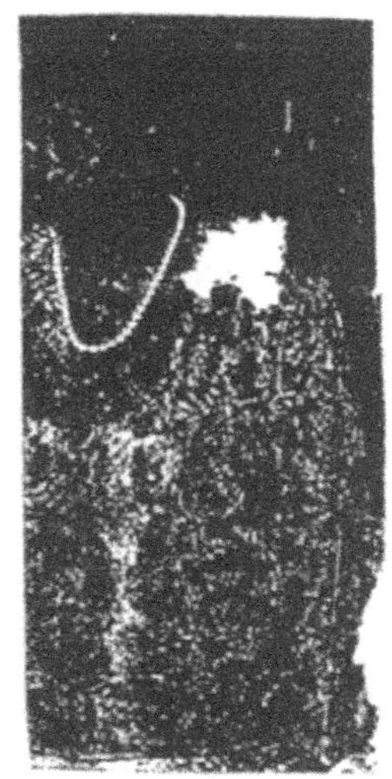

Eddie Louis

Evelyn

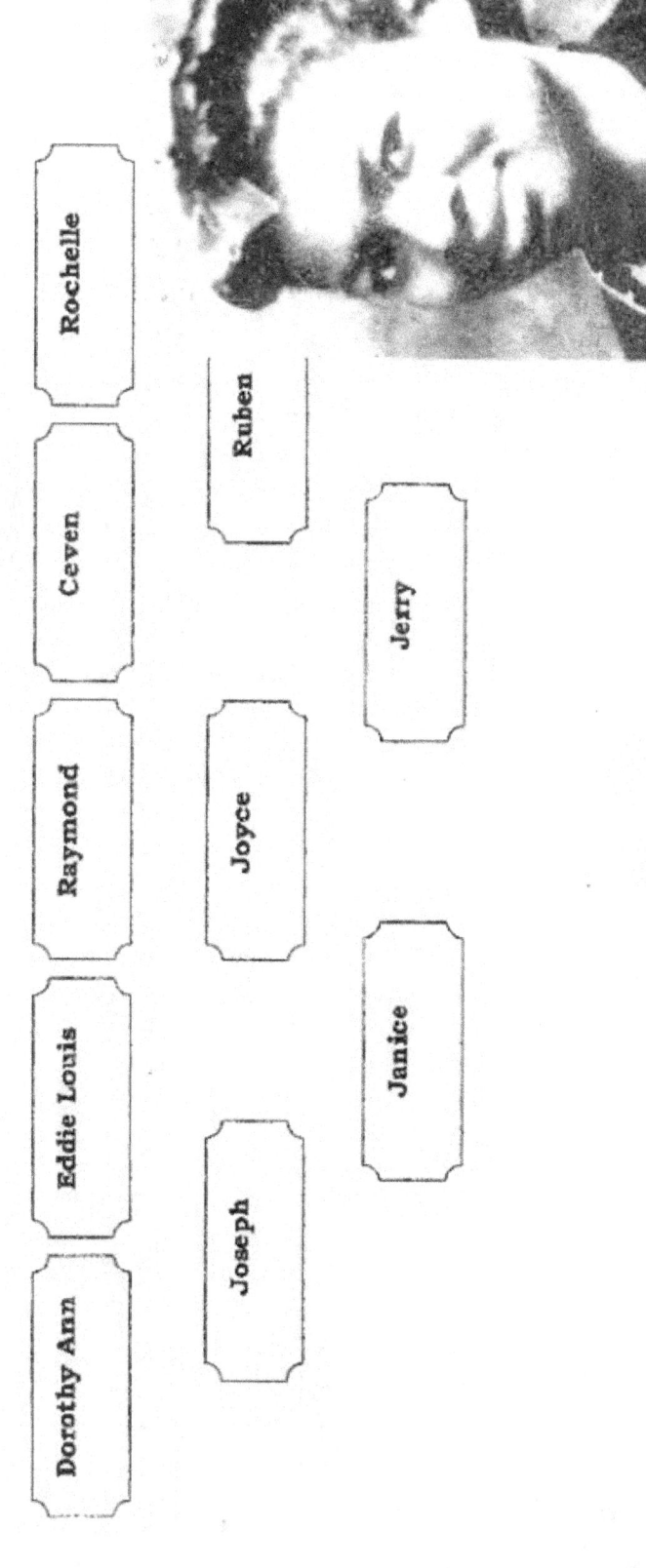

Favoring her mother Evelyn had a good sense of humor. She worked for Dr. Beasly in Greenwood before following in the footsteps of her siblings and moving to New York. Shortly after arriving in New York Evelyn married, Jim Hunter. Unfortunately, Evelyn died at a very early age. Her descendents are as follow:

Dorothy Ann | **Eddie Louis** | **Raymond** | **Ceven** | **Rochelle**

Joseph | **Joyce** | **Ruben**

Janice | **Jerry**

Edd

With similar features as his brother Pop and a no one sense attitude like Willie Mae, Edd did not play and did not take any static. The Patterson children grew up in a mean south in which Segregation was still very prevalent. Similar too many black men during this time Edd was not afraid of any one (including our white brother and sisters). Refusing to be undermined or belittled Edd spoke his mind. Because of his boldness, the south did not take kindly to him, which resulted in his migration to the north to be with his siblings. Settling in Brooklyn, New York Edd never married, but worked hard in construction until his death. Edd, like some of his brothers was a quiet man. He died of a heart attack in his early sixty's. He is buried in the same commentary as his sister, Carry in New Jersey. Edd never married nor had children.

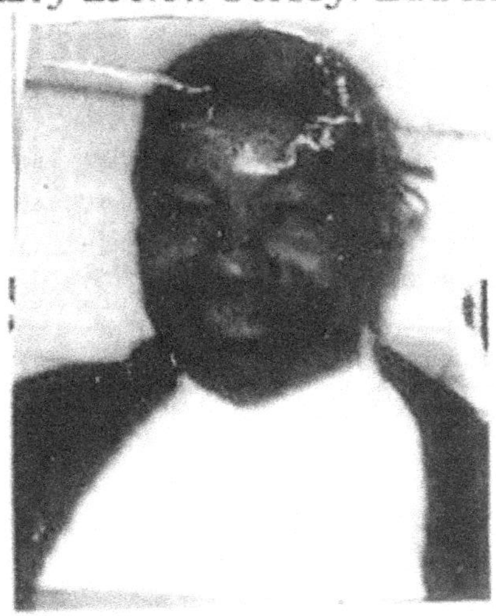

Cora Emma (Sis)

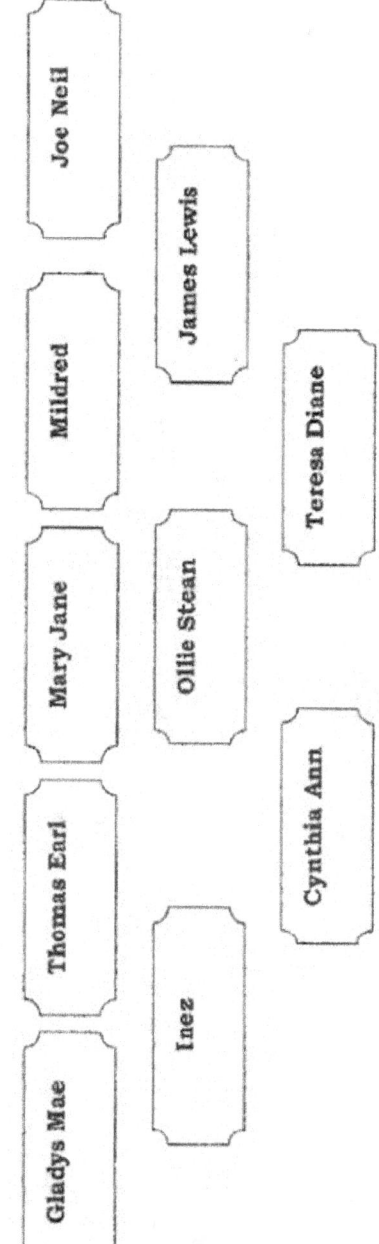

Sis-Sis-Sis! She was great fun to be around. She was the tallest girl and walked with authority. She was the baby of the Patterson clan and often times used it to her advantage. Sis would go toe to toe with her siblings. She was not afraid. Sis had limited education. She met and fell in love with Willie James Gant. They moved to Greenville, SC and raised ten children. She loved her family and enjoyed journeying back home to spend time with her brothers and sisters, especially, Willie Mae. Sis worked as a cook in a school cafeteria until she retired. She loved the Lord and served her Church diligently on the choir and missionary until He called her home on May 28, 1999. Cora was the last of her siblings to die. Their descendents are as follow:

Gladys Mae | **Thomas Earl** | **Mary Jane** | **Mildred** | **Joe Neil**

Inez | **Ollie Stean** | **James Lewis**

Cynthia Ann | **Teresa Diane**

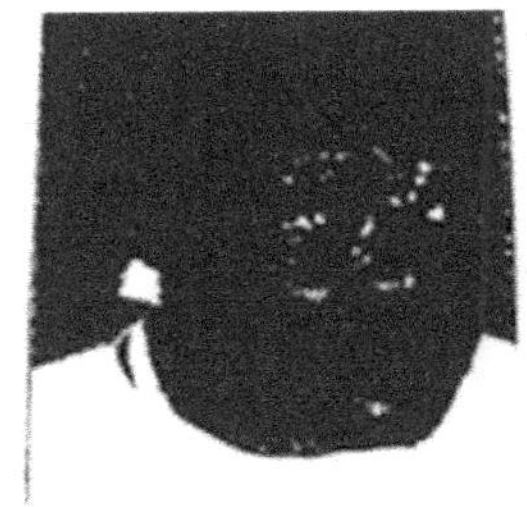

Kathleen "Kat"

Kathleen "Kat" was a very lively person. She was tall and medium build. She was the first grand child of Ruben and Willie Ann. She was a great joy to have in the house. Ruben and Willie Ann raised Kathleen. She grew up more of a sister than a niece. As a young lady Kat worked for the Beasley's doing housework. Kathleen did domestic work for most of her life. She met and fell in love with Thomas Brooks. They married and began their life as one. For a brief time they moved to Brooklyn, New York. They return south to Greenwood where they had their four children. Kathleen loved to cook, especially her cakes, pies and yeast rolls. Like her sister, Willie Mae she cooked everything from scratch. Kathleen loved her family and enjoyed being around them. She also loved the LORD and proved so through her work in the Church, Beulah Baptist. She served as the Church Secretary, sung on the choir and missionary. She loved to laugh and sing. She would have a distinctive laugh and it could be heard throughout the house. Kathleen went home to be with the LORD in her seventy's.

Barrett | **Darlene** | **Ernestine** | **Melvin**

The Reuben Patterson Story

The following pages are just the beginning of the story of our family history that I began several months ago for our benefit and for our generations that follow us. To be successful, I desperately need your help by sending in information to record. I would like for each of you to take these pages, study them, then collect your own family info and send to me. As you collect information, remember the older family members. They are carrying around a wealth of info in their heads, never written down. Some folks say that the first time ancestors are mentioned to an older person, they won't remember a thing. But somehow this makes them start thinking about their ancestors they remember and by the next time you contact them, they are eager to talk. You might want to take a tape recorder with you so you can save the sound of their voices and their knowledge.

Information needed is a listing of births, deaths, marriages, where born, where died, where married, maiden name of wife, surname of husband, just any info you have. List the parents with their children, the children's children and do this for all the children for each set of parents. Go as far back as possible, as you may just provide the clue to older kinfolks that we don't know about. Give the name of the county or town, or both, along with the state, as our family did travel around and lived different places. The place of burial, like the name of the church or cemetery, will also be useful – just anything you can think of about our family to help fill in the gaps. Family tradition among your families will be nice. This might not be 100% right, but normally most of it is correct. At least it will be written down for others to share and we can work to prove it. We are eager to find more family members of Willis Patterson, Reuben's father and on Nancy McKenseley, Reubin's mother. We know nothing about her, so send anything you have.

Don't forget to make corrections if you see something wrong in these pages. We do want to get it right. All this will ensure a much larger, more informative booklet to be distributed next year at our reunion.

I am now collecting Holloway information, so please contribute to this project also. To make filing easier, I ask you to put the Holloway info on separate pages so I won't have to recopy. Hopefully next year we will also have a Holloway booklet.

I would like to thank the following persons for their help:

Steve Tuttle, S. C. Dept. of Archives and History
Barbara Segler, Georgia Dept. of Archives
Tonya Taylor, Edgefield Tompkins Library, Searchroom
Peggy Brakefield, Camden Archives
Wilma Kirkland, who says she has thoroughly enjoyed the challenging experience.

Send information by mail to: Dorothy Chappell (Mrs. Timothy)
4426 Cokesbury Rd.
Hodges, SC 29653
Send information by email to: Cynthia Chappell
Bbs1028@wmconnect.com

National Park Service
U.S. Department of the Interior

Charles Pinckney National Historic Site
Mt. Pleasant, South Carolina

African Americans at Snee Farm

Topic Page | Feedback | Search Pages | Learn More | Create Page

Slaves at Snee Farm Plantation, Charleston, SC, 1859

317 views. Created by Lowcountry_Africana.

This page honors 62 enslaved ancestors listed at Snee Farm Plantation in the 1859 estate inventory of William McCants of Charleston, South Carolina. Anyone may contribute to this page.

Photos (4)

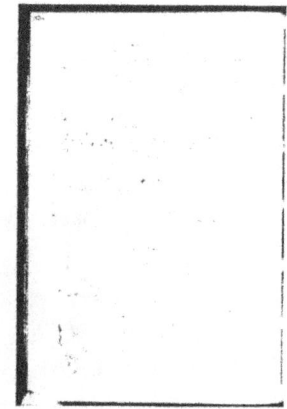

Search for images on Footnote matching **Slaves at Snee Farm Plantation, Charleston, SC, 1859**

Places mentioned on this page

Timeline

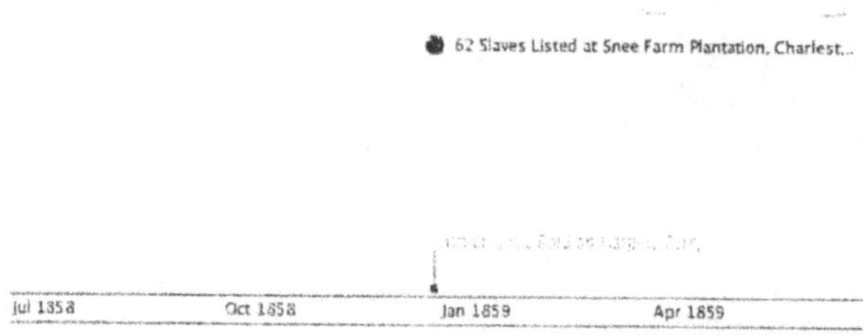

Facts

There are no facts.

Stories

62 Slaves Listed at Snee Farm Plantation, Charleston, SC, 1859

1859 | Snee Farm Plantation, Charleston, SC

This page honors 62 enslaved ancestors listed at Snee Farm Plantation in the 1859 estate inventory of William McCants of Charleston, South Carolina.

Anyone may contribute to this page.

To view the document, please click on the images attached to this page.

Slaves listed at Snee Farm were:

- Adam – slave
- Nancy – slave
- Infant – slave
- Bess – slave
- Jimmy – slave
- Lucy – slave
- Minda – slave
- Moses – slave
- Louisa – slave
- Stephen – slave
- Tyrah – slave
- Phoeby – slave
- WarC – slave
- Nora – slave
- Joe – slave
- William – slave
- Abram – slave
- Dido – slave
- Paul – slave
- Aron – slave
- Sally – slave
- Hester – slave
- Little Aron – slave
- Celey – slave
- Betty – slave
- Robert – slave
- Jane – slave
- Judy – slave
- Sarah – slave

Connected Pages

Lowcountry Africana: South Carolina Slave Records on Footnote.com
See all of the pages we've built for SC Estate Inventories.

Lowcountry Africana on Footnote.com
This page was created by Lowcountry Africana.

Restore the Ancestors Indexing Project: SC Estate Inventories
You can volunteer to annotate these documents to make them searchable.

Links

There are no links about Slaves at Snee Farm Plantation, Charleston, SC, 1859.

About this page

Anyone can contribute to this page.

Original author: Lowcountry_Africana
Created Date: 11 Jul 2010
Page views: 317 total (20 this week)

Flag this page for abuse

- Katy - slave
- James - slave
- Prince - slave
- Daphne - slave
- Jacob - slave
- Jeffery - slave
- Isaac - slave
- Maria - slave
- Hagar - slave
- Trumpeter - slave
- Philis - slave
- David - slave
- Jane - slave
- Patty - slave
- Amey - slave
- Neptune - slave
- Auber - slave
- Richard - slave
- Daniel - slave
- Clementine - slave
- Samuel - slave
- Charlotte - slave
- Lovey - slave
- Elsey - slave
- Amey - slave
- Old Phillis - slave
- Nancy - slave
- Frederick - slave
- Hannah - slave
- Sam - slave
- William - slave
- Kate - slave

Created by Lowcountry_Africana 11 Jul 2010

About This Document

This document was digitized as part of a collaborative effort between Footnote, Lowcountry Africana, the South Carolina Department of Archives and History and FamilySearch, to digitize all surviving Charleston, SC Estate Inventories, 1732-1872 and Bills of Sale, 1773-1872, for a FREE Footnote collection.

When the project is complete, the names of more than 30,000 enslaved ancestors will be preserved in this FREE collection, for generations to come.

This remarkable collection contains the name of every slave ever listed in a surviving estate inventory for Charleston, SC from colonial times to Emancipation. Many of the post-Civil War inventories list the names of former slaves as well. You can volunteer to help index this FREE collection, to make the records searchable.

To learn more please CLICK HERE.

This document was indexed and made searchable by Alana Thevenet.

Created by Lowcountry_Africana 19 Jul 2010

Comments

There are no comments.

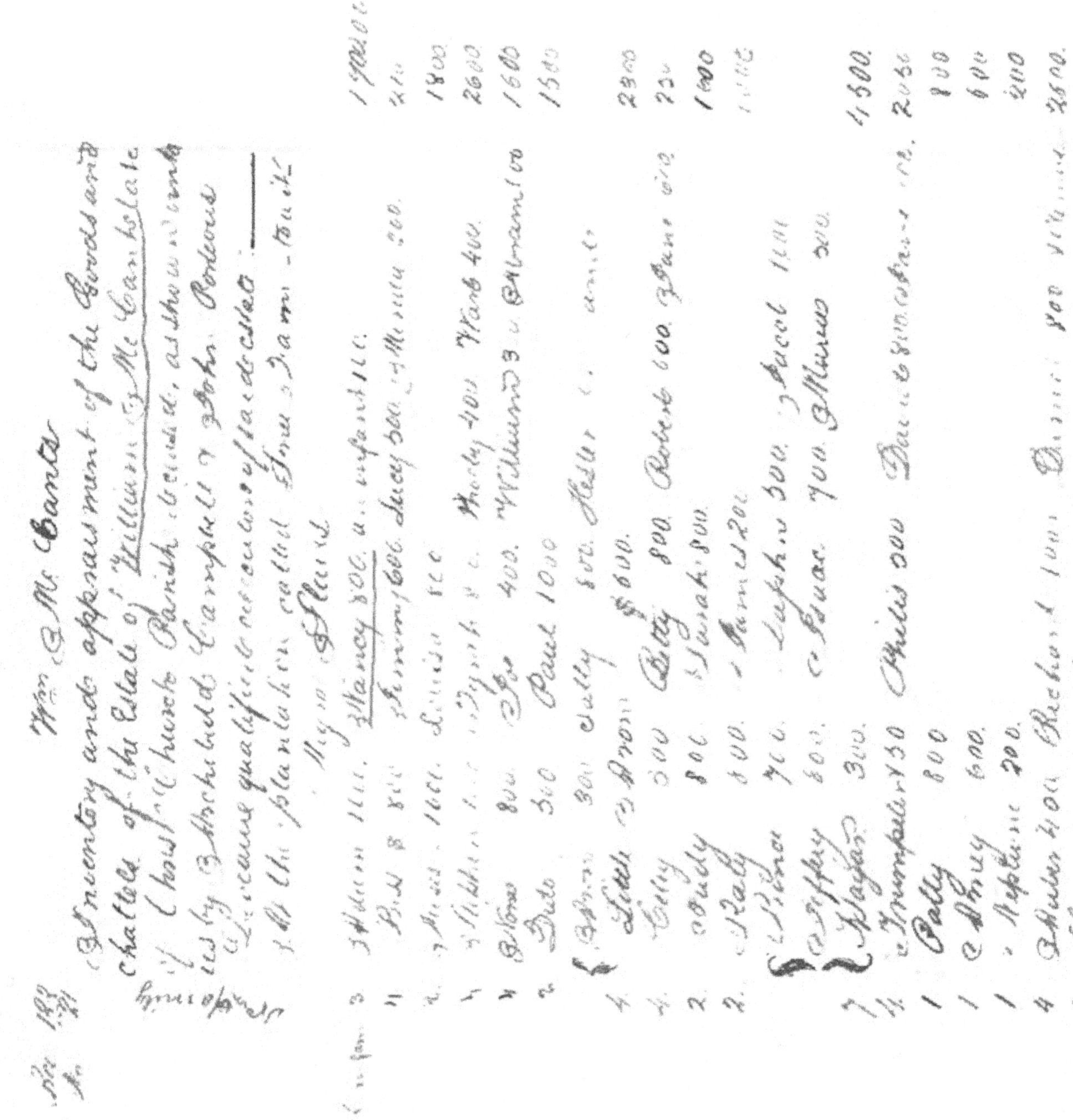

Rec 183 An 21

Wm. G. McCants

Inventory and appraisement of the Goods and chattels of the Estate of William G. McCants late of Christ Church Parish deceased, as shown unto us by Archibald Campbell & John Porteous became qualified executors of said estate: —

At the plantation called [illegible]

Negro Slaves

3	Adam 1000. Nancy 800. an infant 100.	1900.00
4	Bill 800 Jimmy 600. Lucy 500. Minda 200.	[illegible]
2	Moses 1000. Louisa 800	1800
4	Hibben [illegible] Dinah [illegible] Phoeby 400 Ward 400.	2600
4	Nero 800 Joe 400. William 300 Abram 100	1600
2	Dido 500 Paul 1000	1500
4	Aron 300 Sally 800. Hester [illegible] and Little Aron $600.	2300
4	Cely 500 Betty 800. Robert 600. June 600.	250[illegible]
2	Judy 800 Sarah 800.	1600
2	Katy 800. James 200	1000
7	Prince 700. Lafslin 500. Jacob 1000 Jeffery 800. Isaac 700. Marcus 500. Hagar 300.	4500.
4	Trumpeter 750 Philis 500 David 800 [illegible]	2050
1	Patty 800	800
1	Amey 600.	600
1	Neptune 200.	200
4	Hester 400 Richard 1000 Daniel 800 [illegible]	2800.
2	Samuel 1000 Charlotte [illegible]	18[illegible]

SLAVERY IN THE SOUTH

AND SOUTH CAROLINA

'Taint no use of me working so hard-

I've got a gal in the white folks yard,

When she kills a chicken she saves me the head

She thinks I'm working but I'm lying in the bed.

Who stole a lock? I don't know.

Who stole a lock from the hen house door?

I'm gonna find out before I go,

Who stole a lock from the hen house door.

'Taint no use of me working so hard,

I've got a gal in the white folks yard,

When she kills a chicken she saves me the wing

She thinks I'm working but I ain't doing a thing.

Taint no use of me working so hard,

I've got a gal in the white folks yard,

When she kills a chicken she saves me the feet

She thinks I'm working but I'm loafing the street.

Close to two million slaves were brought to the American South from Africa and the West Indies during the centuries of the Atlantic slave trade. Approximately 20% ofthe population of the American South over the years has been African American, and as late as 1900, 9 out of every 10 African Americans lived in the South. The large number of black people maintained as a labor force in the post-slavery South were not permitted to threaten the region's character as a white man's country, however. The region's ruling class dedicated itself to the overriding principle ofwhite supremacy, and white racism became the driving force of southern race relations. The culture of racism sanctioned and supported the whole range of discrimination that has characterized white supremacy in its successive stages. During and after the slavery era, the culture of white racism sanctioned not only official systems of discrimination but a complex code of speech, behavior, and social practices designed to make white supremacy seem not only legitimate but natural and inevitable.

In the antebellum South, slavery provided the economic foundation that supported the dominant planter ruling class. Under slavery the structure of white supremacy was hierarchical and patriarchal, resting on male privilege and masculinist honor, entrenched economic power, and raw force. Black people necessarily developed their sense of identity, family relations, communal values, religion, and to an impressive extent their cultural autonomy by exploiting contradictions and opportunities within a complex fabric of paternalistic give-and-take. The working relationships and sometimes tacit expectations and obligations between slave and slave holder made possible a functional, and in some cases highly profitable, economic system.
Academic Affairs Library, The University ofNorth Carolina at Chapel Hill

Beginnings

Slavery played a central role in the history ofthe United States. It existed in all the English mainland colonies and came to dominate agricultural production in the states from Maryland south. Eight of the first 12 presidents ofthe United States were slave owners. Debate over slavery increasingly dominated American politics, leading eventually to the American Civil War, which finally brought slavery to an end. After emancipation, overcoming slavery's legacy remained a crucial issue in American history, from Reconstruction following the war, to the civil rights movement a century later.

Slavery has appeared throughout history in many forms and many places. Slaves have served in capacities as diverse as concubines, warriors, servants, craft workers, and tutors. In the Americas, however, slavery emerged as a system of forced labor designed for the production of staple crops. Depending on location, these crops included sugar, tobacco, coffee, and cotton :in the southern United States, by far the most important staples were tobacco and cotton.

South Carolina is about the same size as Sierra Leone and has a roughly similar geography and climate. There is the "Low Country" which consists of the Sea Islands, the swampy southern coastline, and a wide and fertile arc of coastal plain stretching up to a hundred miles in the interior. Beyond that is the "Upcountry," a region of rolling hills rising gradually to mountains three thousand feet high in the far northwest. Much of the state is humid and semitropical with long, hot summers and mild winters and abundant rainfall reaching seventy inches in some areas. Three-fifths of the state is covered in forest, and a series of rivers flows down in parallel lines to the Atlantic coast.

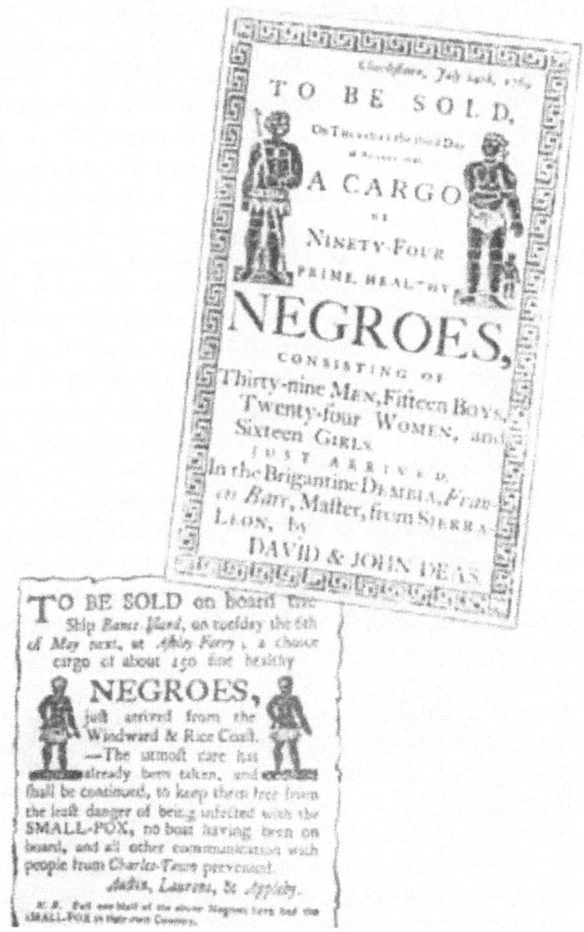

CAPTION: South Carolina planters were willing to pay higher prices for slaves from rice-growing areas. Upper photo-slaves from "Sierra-Leon"; Lower photo-slaves from the "Windward and Rice Coast" arriving on the ship Bance Island. Henry Laurens' name appears at bottom.

The first English-speaking settlement in South Carolina was established on the coast in 1670. For the first thirty years the colonists had little success, but by about 1700 they discovered that rice, imported from Asia, grew well in the inland valley swamps of the Low Country. Throughout the 1700s the economy of South Carolina was based overwhelmingly on the cultivation of rice. This product brought consistently high prices in England, and the colony prospered and expanded. Rice agriculture has been called "the best opportunity for industrial profit which 18th century America afforded." South Carolina became one of the richest of the North American Colonies; and Charlestown (now Charleston), its capital and principal port, one of the wealthiest and most fashionable cities in early America. Later, because of the extraordinary success in South Carolina, the rice plantation system was extended farther south into coastal Georgia, where it also prospered.

The South Carolina planters were, at first, completely ignorant of rice cultivation, and their early experiments with this specialized type of tropical agriculture were mostly failures. They soon recognized the advantage of importing slaves from the traditional rice-growing region of West Africa, and they generally showed far greater interest in the geographical origins of African slaves than did planters in other North American colonies. The South Carolina rice planters were willing to pay higher prices for slaves from the "Rice Coast," the Windward Coast," the "Gambia," and "Sierra-Leon"; and slave traders in Africa soon learned that South Carolina was an especially profitable market for slaves from those areas. When slave traders arrived in Charlestown with slaves from the rice-growing region, they were careful to advertise their origin on auction posters or in newspaper announcements, sometimes noting that the slaves were "accustomed to the planting of rice." Traders who arrived in Charlestown with slaves from other parts of Africa where rice was not traditionally grown, such as Nigeria, often found that their slaves fetched lower prices. In some cases, they could sell no slaves at all and had to sail away to another port.

The South Carolina and Georgia colonists ultimately adopted a system of rice cultivation that drew heavily on the labor patterns and technical knowledge of their African slaves. During the growing season the slaves on the rice plantations moved through the fields in a line, hoeing rhythmically and singing work songs to keep in unison. At harvest time the women processed the rice by pounding it in large wooden mortars and pestles, virtually identical to those used in West Africa, and then "fanning" the rice in large round winnowing baskets to separate the grain and chaff. The slaves may also have contributed to the system of sluices, banks, and ditches used on the South Carolina and Georgia rice plantations. West African farmers traditionally cultivated local varieties of wet rice on the flood plains and dry rice on the hillsides. During the 1500s the Portuguese introduced superior types of paddy rice from Asia, and travellers in the 1700s noted that Western African farmers-including the Temne of Sierra Leone-were constructing elaborate irrigation systems for rice cultivation. In South Carolina and Georgia the slaves simply continued with many of the methods of rice farming to which they were accustomed in Africa.

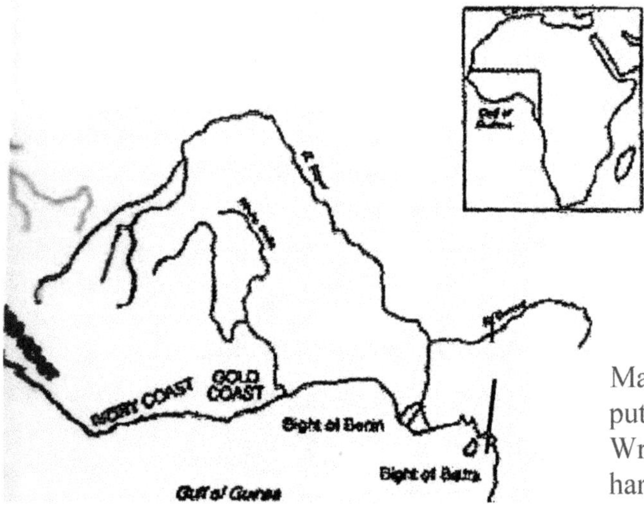

West Africa in the era of the slave trade

Carolina planters developed a vision ofthe "ideal" slave-- tall, healthy, male, between the ages of 14and 18, "free of blemishes," and as dark as possible. For these ideal slaves Carolina planters in the eighteenth century paid, on average, between ? 100and ?200 sterling- in today's money that is between $11,630 and $23,200!

Many ofthese slaves were almost immediately put to work in South Carolina's rice fields. Writers ofthe period remarked that there was no harder, or more unhealthy, work possible: negroes, anckle and even mid-leg deep in water which floats an ouzy mud, and exposed all the while to a burning sun which makes the very air they breathe hotter than the human blood; these poor retches are then in a furness of stinking putrid effluvia: a more horrible employment can hardly be imagined.

In fact, these Carolina rice fields have been described as charnel houses for African-American slaves. Malaria and enteric diseases killed off the low country slaves at rates which are today almost unbelievable. Based on the best plantation accounts it is clear that while about one out of every three on the cotton plantations died before reaching the age of 16,nearly two out of every three American children on rice plantations failed to reach their sixteenth birthday and over a third of children died before their first birthday. Rice's macabre record of slave deaths has been traced to two primary factors - one was malaria, the other was the infants' feebleness at birth, probably the result of the mother's own chronic malaria and their general exhaustion from rice cultivation during pregnancy.

After their horrifice "Middle Passage," over 40% ofthe African slaves reaching the British colonies before the American Revolution passed through South Carolina. Almost all ofthese slaves entered the Charleston port, being briefly quarantined on Sullivan's Island, before being sold in Charleston's slave market.

Once in South Carolina what was the lives ofthese slaves like? How did they live? What did they eat? What did their houses look like? How did they prepare their food? What kinds ofpossessions did they have? What did their pottery look like? White masters had little or no interest in recording these details for future generations. Slavery was an economic issue and the only details worthy ofbeing consistently recorded were those related to the value oftheir slaves or the value oftheir production . The daily lives of these new African-Americans was probably poorly understood and certainly of little importance to the planters. These are all questions that can only be answered through archaeology.

Most of the agriculture in the southern United States during the early 19th century was dedicated to growing one crop-cotton. Most of the cotton crop was grown on large plantations that used black slave labor, such as this one on the Mississippi River.

There was nothing inevitable about the use of black slaves. Although 20 Africans were purchased in Jamestown, Virginia, as early as 1619, throughout most of the 17th century the number of Africans inthe English mainland colonies (American Colonies) grew slowly. During those years, colonists experimented with two other sources of forced labor: Native American slaves and European indentured servants. The number ofNative American slaves was limited in part because the Native Americans were in their homeland; they knew the terrain and could escape fairly easily. Although some Native American slaves existed in every colony the number was limited. The settlers found it easier to sell Native Americans captured in war to planters in the Caribbean than to turn them into slaves on their own terrain.

More important as a form of labor was indentured servitude. Most indentured servants were poor Europeans who wanted to escape harsh conditions and take advantage ofopportunities in America. They traded four to seven years of their labor in exchange for the transatlantic passage. At first, indentured servants came mainly from England, but later they came increasingly from Ireland, Wales, and Germany. They were primarily, although not exclusively, young males. Once in the colonies, they were essentially temporary slaves; most served as agricultural workers although some, especially in the North, were taught skilled trades. During the 17th century, they performed most ofthe heavy labor in the Southern colonies and also provided the bulk of immigrants to those colonies.

Ex-slaves sitting in front of a cabin. This picture is from the main eastern theater of war, The Peninsular Campaign, May-August 1862.

Slave Trade

Because the labor needs of the rapidly growing colonies were increasing, this decline in servant migration produced a labor crisis. To meet it, landowners turned to African slaves, who from the 1680s began to replace indentured servants; in Virginia, for example, blacks, the great majority of whom were slaves, increased from about 7 percent of the population in 1680 to more than 40 percent by the mid-18th century. During the first half of the 17th century, the Netherlands and Portugal had dominated the African slave trade and the number of Africans available to English colonists was limited because the three countries competed for slave labor to produce crops in their American colonies. During the late 17th and 18th centuries, by contrast, naval superiority gave England a do.minant position in the slave trade, and English traders transported millions of Africans across the Atlantic Ocean.

Since others died before boarding the ships, Africa's loss of population was even greater. By far the largest importers of slaves were Brazil and the Caribbean colonies; together, they received more than three-quarters of all Africans brought to the Americas. About 6 percent of the total (600,000 to 650,000 people) came to what is now the United States.

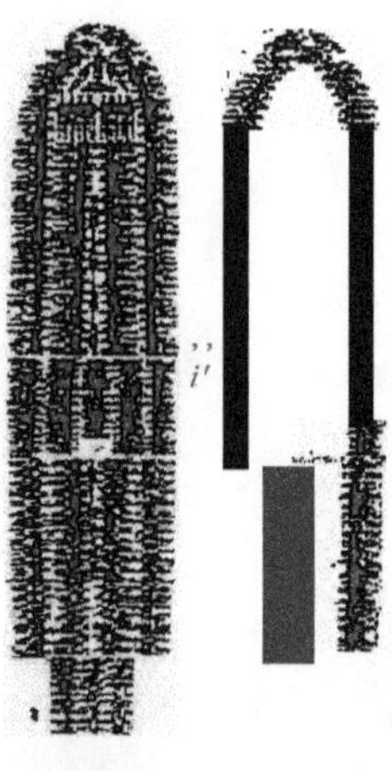

The transatlantic slave trade produced one of the largest forced migrations in history. From the early 16th to the mid-19th centuries, between 10 million and 11 million Africans were taken from their homes, herded onto ships where they were sometimes so tightly packed that they could barely move, and sent to a strange new land.

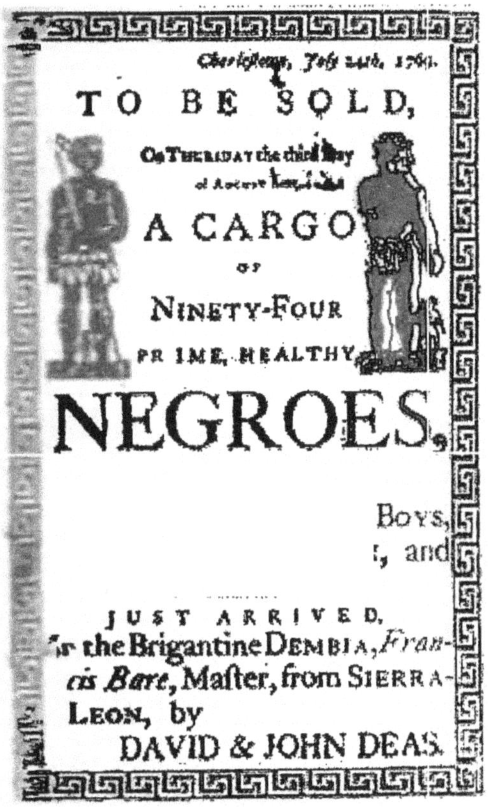

In 1769 the firm of David & John Deas advertised
The sale of 94 African-Americans in Charleston, SC

-Spread of Slavery-

Slavesperformed numerous tasks, from clearing forests to serving as guides, trappers, craft workers, nurses, and house servants, but they were most essential as agricultural laborers. Slaves were most numerous where landowners sought to grow staple crops for market, such as tobacco in the upper South (Maryland, Virginia, North Carolina) and rice in the lower South (South Carolina, Georgia). Slaves also worked on large wheat-producing estates in New York and on horse-breeding farms in Rhode Island, but climate and soil restricted the development of commercial agriculture in the Northern colonies, and slavery never became as economically important as it did in the South.

nation.

Introduction

Slavery in America began when the first African slaves were brought to the North American colony of Jamestown, Virginia, in 1619, to aid in the production of such lucrative crops as tobacco. Slavery was practiced throughout the American colonies in the 17th and 18th centuries, and African-American slaves helped build the economic foundations of the new nation. The invention of the cotton gin in 1793 solidified the central importance of slavery to the South's economy. By the mid-19th century, America's westward expansion, along with a growing abolition movement in the North, would provoke a great debate over slavery that would tear the nation apart in the bloody American Civil War (1861-65). Though the Union victory freed the nation's 4 million slaves, the legacy of slavery continued to influence American history, from the tumultuous years of Reconstruction (1865-77) to the civil rights movement that emerged in the 1960s, a century after emancipation.

Harriet Tubman and the Underground Railroad

Frederick Douglass

Foundations of Slavery in America

In the early 17th century, European settlers in North America turned to African slaves as a cheaper, more plentiful labor source than indentured servants (who were mostly poorer Europeans). After 1619, when a Dutch ship brought 20 Africans ashore at the British colony of Jamestown, Virginia, slavery spread throughout the American colonies. Though it is impossible to give accurate figures, some historians have estimated that 6 to 7 million slaves were imported to the New World during the 18th century alone, depriving the African continent of some of its healthiest and ablest men and women.

Did You Know?

One of the first martyrs to the cause of American patriotism was Crispus Attucks, a former slave who was killed by British soldiers during the Boston Massacre of 1770. Some 5,000 black soldiers and sailors fought on the American side during the Revolutionary War.

In the 17th and 18th centuries, black slaves worked mainly on the tobacco, rice and indigo plantations of the southern coast. After the American Revolution (1775-83).many colonists (particularly in the North. where slavery was relatively unimportant to the economy) began to link the oppression of black slaves to their own oppression by the British, and to call for slavery's abolition. After the war's end, however, the new U.S. Constitution tacitly acknowledged the institution, counting each slave as three-fifths of a person for the purposes of taxation and representation in Congress and guaranteeing the right to repossess any "person held to service or labor" (an obvious euphemism for slavery).

Importance of the Cotton Gin

In the late 18th century, with the land used to grow tobacco nearly exhausted, the South faced an economic crisis, and the continued growth of slavery in America seemed in doubt. Around the same time, the mechanization of the textile industry in England led to a huge demand for American cotton, a southern crop whose production was unfortunately limited by the difficulty of removing the seeds from raw cotton fibers by hand. In 1793, a young Yankee schoolteacher named Eli Whitney invented the cotton gin, a simple mechanized device that efficiently removed the seeds. His device was widely copied, and within a few years the South would transition from the large-scale production of tobacco to that of cotton, a switch that reinforced the region's dependence on slave labor.

Slavery itself was never widespread in the North, though many of the region's businessmen grew rich on the slave trade and investments in southern plantations. Between 1774 and 1804, all of the northern states abolished slavery, but the so-called "peculiar institution" remained absolutely vital to the South. Though the U.S. Congress outlawed the African slave trade in 1808, the domestic trade flourished, and the slave population in the U.S. nearly tripled over the next 50 years. By 1860 it had reached nearly 4 million, with more than half living in the cotton-producing states of the South.

Slaves and Slaveholders

Slaves in the antebellum South constituted about one-third of the southern population. Most slaves lived on large farms or small plantations; many masters owned less than 50 slaves. Slave owners sought to make their slaves completely dependent on them, and a system ofrestrictive codes governed life among slaves. They were prohibited from learning to read and write, and their behavior and movement was restricted. Many masters took sexual liberties with slave women, and rewarded obedient slave behavior with favors, while rebellious slaves were brutally punished. A strict hierarchy among slaves (from privileged house slaves and skilled artisans down to lowly field hands) helped keep them divided and less likely to organize against their masters. Slave marriages had no legal basis, but slaves did marry and raise large families; most slave owners encouraged this practice, but nonetheless did not hesitate to divide slave families by sale or removal.

Slave revolts did occur within the system (notably ones led by Gabriel Prosser in Richmond in 1800 and by Denmark Vesey in Charleston in 1822), but few were successful. The slave revolt that most terrified white slaveholders was that led by Nat Turner in Southampton County, Virginia, in August 1931. Turner's group, which eventually numbered around 75 blacks, murdered some 60 whites in two days before armed resistance from local whites and the arrival of state militia forces overwhelmed them. Supporters of slavery pointed to Turner's rebellion as evidence that blacks were inherently inferior barbarians requiring an institution such as slavery to discipline them, and fears of similar insurrections led many southern states to further strengthen their slave codes in order to limit the education, movement and assembly of slaves. In the North, the increased repression of southern blacks would only fan the flames of the growing abolition movement.

Rise of the Abolition Movement

From the 1830s to the 1860s, a movement to abolish slavery in America gained strength in the northern United States, led by free blacks such as Frederick Douglass and white supporters such as William Lloyd Garrison, founder of the radical newspaper The Liberator, and Harriet Beecher Stowe, who published the bestselling antislavery novel "Uncle Tom's Cabin" (1852). While many abolitionists based their activism on the belief that slaveholding was a sin, others were more inclined to the non-religious "free-labor" argument, which held that slaveholding was regressive, inefficient and made little economic sense.

Free blacks and other antislavery northerners had begun helping fugitive slaves escape from southern plantations to the North via a loose network of safe houses as early as the 1780s. This practice, known as the Underground Railroad, gained real momentum in the 1830s and although estimates vary widely, it may have helped anywhere from 40,000 to 100,000 slaves reach freedom. The success of the Underground Railroad helped spread abolitionist feelings in the North; it also undoubtedly increased sectional tensions, convincing pro-slavery southerners of their northern countrymen's determination to defeat the institution that sustained them.

Western Expansion and Debate over Slavery in America

America's explosive growth-and its expansion westward in the first half of the 19th century-would provide a larger stage for the growing conflict over slavery in America and its future limitation or expansion. In 1820, a bitter debate over the federal government's right to restrict slavery over Missouri's application for statehood ended in a compromise: Missouri was admitted to the Union as a slave state, Maine as a free state and all western territories north of Missouri's southern border were to be free soil. Although the Missouri Compromise was designed to maintain an even balance between slave and free states, it was able to help quell the forces of sectionalism only temporarily.

In 1850, another tenuous compromise was negotiated to resolve the question of territory won during the Mexican War. Four years later, however, the Kansas-Nebraska Act opened all new territories to slavery by asserting the rule of popular sovereignty over congressional edict, leading pro- and anti-slavery forces to battle it out (with much bloodshed) in the new state of Kansas. Outrage in the North over the Kansas-Nebraska Act spelled the downfall of the old Whig Party and the birth of a new, all-northern Republican Party. In 1857, the Supreme Court's ruling in the Dred Scott case (involving a slave who sued for his freedom on the grounds that his master had taken him into free territory) effectively repealed the Missouri Compromise by ruling that all territories were open to slavery. The abolitionist John Brown's raid at Harper's Ferry, Virginia, in 1859 aroused sectional tensions even further: Executed for his crimes, Brown was hailed as a martyred hero by northern abolitionists and a vile murderer in the South.

Civil War and Emancipation

The South would reach the breaking point the following year, when Republican candidate Abraham Lincoln was elected as president. Within three months, seven southern states had seceded to form the Confederate States of America; four more would follow after the Civil War (1861-65) began. Though Lincoln's antislavery views were well established, the central Union war aim at first was not to abolish slavery, but to preserve the United States as a nation. Abolition became a war aim only later, due to military necessity, growing anti-slavery sentiment in the North and the self-emancipation of many African Americans who fled enslavement as Union troops swept through the South. Five days after the bloody Union victory at Antietam in September 1862, Lincoln issued a preliminary emancipation proclamation, and on January 1, 1863, he made it official that "slaves within any State, or designated part of a State...in rebellion,...shall be then, thenceforward, and forever free."

By freeing some 3 million black slaves in the rebel states, the Emancipation Proclamation deprived the Confederacy of the bulk of its labor forces and put international public opinion strongly on the Union side. Some 186,000 black soldiers would join the Union Army by the time the war ended in 1865, and 38,000 lost their lives. The total number of dead at war's end was 620,000 (out of a population of some 35 million), making it the costliest conflict in American history.

The Legacy of Slavery

The 13th Amendment, adopted late in 1865, officially abolished slavery, but freed blacks' status in the post-war South remained precarious, and significant challenges awaited during the Reconstruction period (1865-77). Former slaves received the rights of citizenship and the "equal protection" of the Constitution in the 14th Amendment (1868) and the right to vote in the 15th (1870), but the provisions of Constitution were often ignored or violated, and it was difficult for former slaves to gain a foothold in the post-war economy thanks to restrictive black codes and regressive contractual arrangements such as sharecropping.

Despite seeing an unprecedented degree of black participation in American political life, Reconstruction was ultimately frustrating for African Americans, and the rebirth of white supremacy-including the rise of racist organizations such as the Ku Klux Klan-had triumphed in the South by 1877. Almost a century later, resistance to the lingering racism and discrimination in America that began during the slavery era would lead to the civil rights movement of the

1960s, which would achieve the greatest political and social gains for blacks since Reconstruction.

In 1868, the 14th Amendment to the Constitution of the United States granted citizenship and equal civil and legal rights to African Americans and slaves who had been emancipated after the American Civil War, including them under the umbrella phrase "all persons born or naturalized in the United States." In all, the amendment comprises five sections, four ofwhich began in 1866 as separate proposals that stalled in legislative process and were amalgamated into a single amendment.

Article XIV.

Section 1. All persons born or naturalized in the United States, and subject to the jurisdiction thereof, are citizens of the United States and of the state wherein they reside. No State shall make or enforce any law which shall abridge the privileges or immunities of citizens of the United States, nor shall any State deprive any person of life, liberty, or property, without due process of law, nor deny to any person within its jurisdiction the equal protection of the laws.

The 15th Amendment, granting African-American men the right to vote, was formally adopted into the U.S. Constitution on March 30, 1870. Passed by Congress the year before, the amendment reads: "the right of citizens of the United States to vote shall not be denied or abridged by the United States or by any State on account ofrace, color, or previous condition of servitude." Despite the amendment, by the late 1870s, various discriminatory practices were used to prevent African Americans from exercising their right to vote, especially in the South. After decades of discrimination, the Voting Rights Act of 1965 aimed to overcome legal barriers at the state and local levels that denied blacks their right to vote under the 15th Amendment.

Fifteenth Amendment

THE 15TH AMENDMENT: RATIFICATION

In 1867, following the American Civil War (1861-65), the Republican-dominated U.S. Congress passed the First Reconstruction Act, over President Andrew Johnson's veto, dividing the South into five military districts and outlining how new governments based on universal manhood suffrage were to be established. With the adoption of the 15th Amendment in 1870, a politically mobilized African-American community joined with white allies in the Southern states to elect the Republican Party to power, which brought about radical changes across the South. By late 1870, all the former Confederate states had been readmitted to the Union, and most were controlled by the Republican Party, thanks to the support of black voters.

Did You Know?

> *One day after it was ratified, Thomas Mundy Peterson (1824-1904) of Perth Amboy, New Jersey, became the first black person to vote under the authority of the 15th Amendment.*

In the same year, Hiram Rhoades Revels (1827-1901), a Republican from Natchez, Mississippi, became the first African American ever to sit in the U.S. Congress, when he was elected to the U.S. Senate. Although black Republicans never obtained political office in proportion to their overwhelming electoral majority, Revels and a dozen other black men served in Congress during Reconstruction, more than 600 served in state legislatures and many more held local offices.

THE 15TH AMENDMENT: POST-RECONSTRUCTION ERA

In the late 1870s, the Southern Republican Party vanished with the end of Reconstruction, and Southern state governments effectively nullified the 14th amendment (passed in 1868, it guaranteed citizenship and all its privileges to African Americans) and the 15th amendment, stripping blacks in the South of the right to vote. In the ensuing decades, various discriminatory practices including poll taxes and literacy tests, along with intimidation and violence, were used to prevent African Americans from exercising their right to vote.

ITHE 15TH AMENDMENT AND THE VOTING RIGHTS ACT OF 1965

The Voting Rights Act, signed into law by President Lyndon Johnson (1908-73) on August 6, 1965, aimed to overcome legal barriers at the state and local levels that denied African Americans their right to vote under the 15th Amendment.

The act banned the use ofliteracy tests, provided for federal oversight of voter registration in areas where less than 50 percent of the nonwhite population had not registered to vote, and authorized the U.S. attorney general to investigate the use of poll taxes in state and local elections (in 1964, the 24th Amendment made poll taxes illegal in federal elections; poll taxes in state elections were banned in 1966 by the U.S. Supreme Court).

After the passage of the Voting Rights Act, state and local enforcement of the law was weak and it often was ignored outright, mainly in the South and in areas where the proportion of blacks in the population was high and their vote threatened the political status quo. Still, the Voting Rights Act gave African-American voters the legal means to challenge voting restrictions and vastly improved voter turnout.

www.ingramcontent.com/pod-product-compliance
Lightning Source LLC
Chambersburg PA
CBHW081233170426
43198CB00017B/2751
* 9 7 8 0 7 8 8 4 5 6 0 0 8 *